A Victorian Handbook of Mechanical Movements

Thomas Walter Barber

Dover Publications, Inc.
Mineola, New York

Bibliographical Note

This Dover edition, first published in 2013, is an unabridged republication of *The Engineer's Sketch-Book of Mechanical Movements, Devices, Appliances, Contrivances, and Details,* originally published in 1890 by E. & F. Spon, London.

Library of Congress Cataloging-in-Publication Data

Barber, Thomas Walter.
[Engineer's sketch-book of mechanical movements, devices, appliances, contrivances, and details]
A Victorian handbook of mechanical movements / Thomas Walter Barber.
p. cm.
Reprint. Originally published: London : E. & F. Spon, 1890, under the title: The Engineer's sketch-book of mechanical movements, devices, appliances, contrivances, and details.
ISBN-13: 978-0-486-49812-6
ISBN-10: 0-486-49812-3
1. Mechanical movements. I. Title.

TJ181.B23 2013
621.8'11—dc23

2012035377

Manufactured in the United States by RR Donnelley
49812304 2018
www.doverpublications.com

PREFACE.

EVERY successful engineer is a born inventor; indeed the daily work of an engineer in practice largely consists in scheming and devising from previous experience new and improved processes, methods, and details for accomplishing them, and for simplifying or cheapening old forms of machinery and the work they produce, to enable him to successfully compete with others, who are perhaps as ingenious and enterprising as himself.

In the work of designing machinery the draughtsman has to rely mainly on his memory for inspiration; and, for lack of an idea, has frequently to wade through numerous volumes to find a detail or movement to effect a particular purpose. Hence, as a rule, every man's work runs in a groove, his productions generally having the stamp of his particular experience and training clearly marked upon them.

In the course of twenty-five years of such experience, I have found the want of such a volume as the present, and endeavoured to supply the deficiency in my own practice by private notes and sketches, gathered promiscuously, until the difficulty of selection and arrangement became so apparent that I began to classify them, as they exist in the following pages. A few weeks of unusual leisure have enabled me to complete this work and amplify it by numerous additions, and it is now presented in the hope that it will be found of equal service to others engaged in the head-splitting, exhausting work of scheming and devising machinery, than which I can conceive of no head-work more wearing and anxious. Several valuable works have already found numerous users, and there is no lack of admirable collections of memoranda, rules, and data for designing and proportioning the various constructive details of machinery; but, as far as I am aware, there is no work in existence which aims at the same purpose as is attempted in the following pages, viz. to provide side by side suggestive sketches of the various methods in use for accomplishing any particular mechanical movement or work, in a form easily referred to, and devoid of needless detail and elaboration. A sketch, properly executed, is—to a practical man—worth a folio of description; and it is to such that these pages are addressed. For the same reason it has been deemed undesirable to add to the various sketches any rules or tables relating to strengths or dimensions, which may be found in numerous well-known volumes.

Any suggestions or additions will be entertained and gratefully acknowledged.

THOMAS WALTER BARBER.

HASTINGS, 1890.

CONTENTS.

A.

B.

C.

Section 1.—ANCHORING.

1. **Rope pulley anchor**—a car which grips by sinking its wheels in the soil; employed for ploughing tackle.
2. **Anchor plate**—buried in the ground below a mass of masonry—for attaching guys, tie rods, &c. Sometimes a frame, or plate, laid on the ground and ballasted, is the method used.
3. **Screw mooring**, screwed into the ground.
4. **Heavy stone** sunk in the ground and having a ring attached; or a mass of concrete, similarly placed, used for guy ropes, tie rods, and foundation bolt attachments.
5. **Grapnel.**
6. **Mushroom anchor.**
7. **Double fluke anchor.**
8. **Martin's patent anchor**, with swivelling flukes. Several other patent anchors are modifications of this.

Stakes, with or without flanges, vertical or horizontal, are sometimes employed, the flanges taking the cross strain of the ties, &c.

Section 2.—ADJUSTING DEVICES.

For adjustment by Screws, see Section 78, and by Wedges, see Section 36. These are the commonest appliances employed. For Cams also, see Section 9. For adjusting Pedestal Brasses, see Section 46.

For adjustments by keys, cotters, &c., see Section 37. See also Nos. 251, 269, and 297.

9. **Split cone sleeves and set screw adjustment** for a revolving standard, or similar detail, where there is much wear or great accuracy is required in the revolving bearing.
10. **Centre-line adjustment** for lathe headstocks, &c.
11. **Variable curve adjustment**; used in compass planes, instruments for drawing arcs of circles, &c.
12. **Vertical shaft footstep adjustment**; employed on millstones, horizontal grinding mills, &c., to regulate the space between the grinding surfaces. See No. 261.

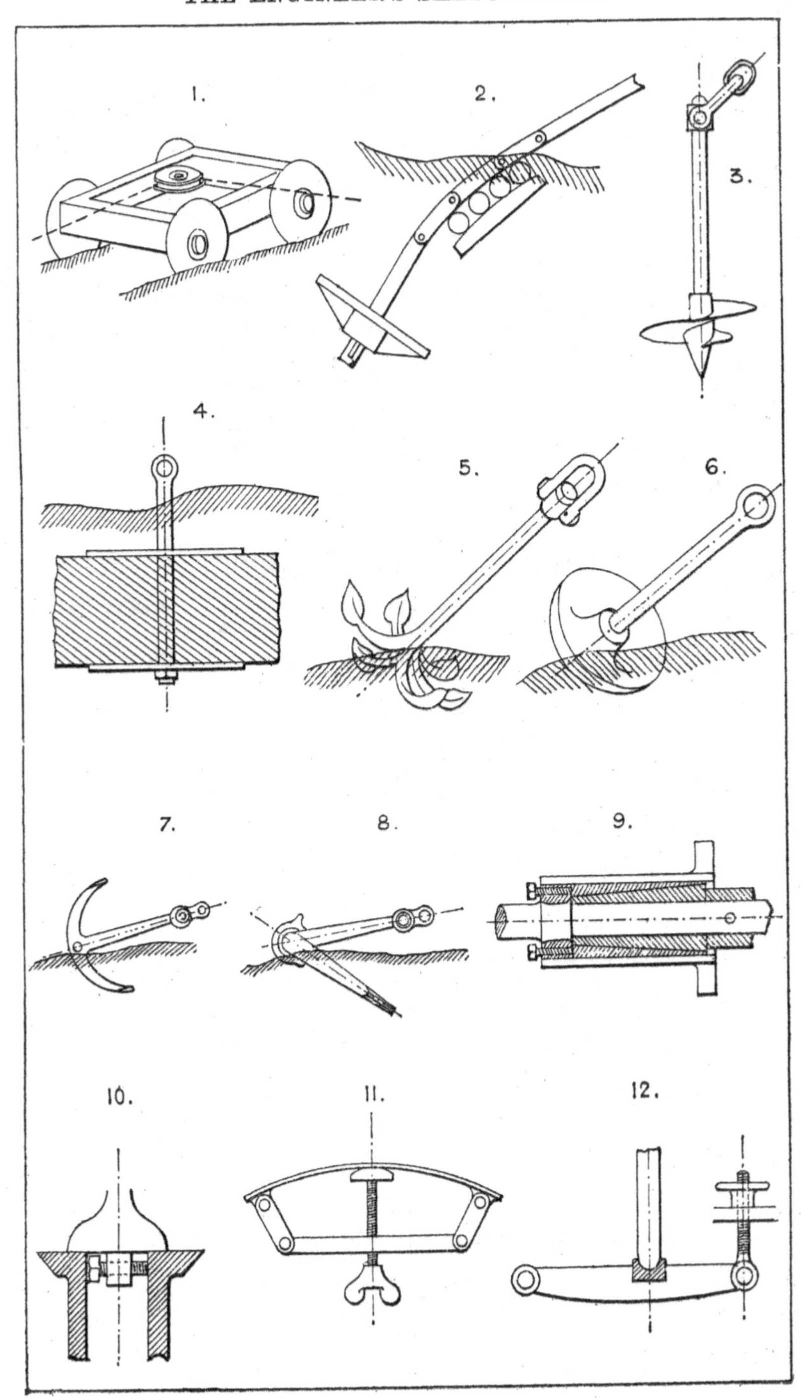
1.
2.
3.
4.
5.
6.
7.
8.
9.
10.
11.
12.

13. **Side screw adjustment** for injectors, jet pumps, &c.
14. **Levelling adjustment**; can be used with either 3 or 4 screws: for telescope and level stands, theodolites, &c.
15. **Horizontal central adjustment** for footsteps, &c.
16. **Slotted link and lock nut** for adjusting angle of a lever.
17. **Disc and ring** with partial angular adjustment by a screw and nut; used for screwing dies, self-centering chucks, &c. The nut and bearing of the screw have allowance for swivelling.
18 **Pin and hole adjustment** for a lever or similar detail.
19. **Wedge bearing** for locomotive horn plate guides, slide bars, and similar parts subject to wear.
20. **Right and left-hand screw and wedge adjustment** for roller bearings, &c.
21. **Adjustment for wear** used on engine crossheads to take up the wear of the working faces.

See also Crane Balance Weights, Section 18.

Section 3.—BELT GEARING.

Materials employed are:—Leather, cotton, guttapercha, indiarubber, canvas, camel-hair, catgut, flat wire or hemp rope, steel bands, flat chains, &c.

22. **Ordinary belt pulley,** "crowned" on face to retain the belt on the centre of the pulley.
23. **Double-flanged pulley,** flat on face, sometimes "crowned," as No. 22.
24. **Single-flanged pulley** for horizontal driving.
25. **Open belt gear**; runs best as shown, with the slack half of the belt at top.
26. **Crossed belt** to reverse motion on the driven shaft.
27. **Mode of driving** when the shafts are at right angles to one another.
28. **Mode of driving** with shafts at an obtuse angle, sometimes used instead of bevel wheels.
29. **Arrangement adopted when the pulleys cannot be got in line** with one another, or the shafts are too close together to drive well direct. Short belts seldom work well.

For reversing by belt gear, see Section 74. Gut bands (round) are worked over V-grooved pulleys; see Rope Gearing, Section 66. Belts may be kept tight by tightening pulleys, see No. 1207. For round belts, see Rope Gearing, Section 66. V-belts are occasionally used, formed of thicknesses of leather riveted together, cut to section, and worked over V-grooved pulleys.

Section 4.—BALL-AND-SOCKET JOINTS.

30. **Universal hinge.** The arm can be fixed in any required position by tightening the gland. Useful for stands for articles to be exhibited in any position, telescopes, &c.
31. **Pipe joint,** with similar capabilities.

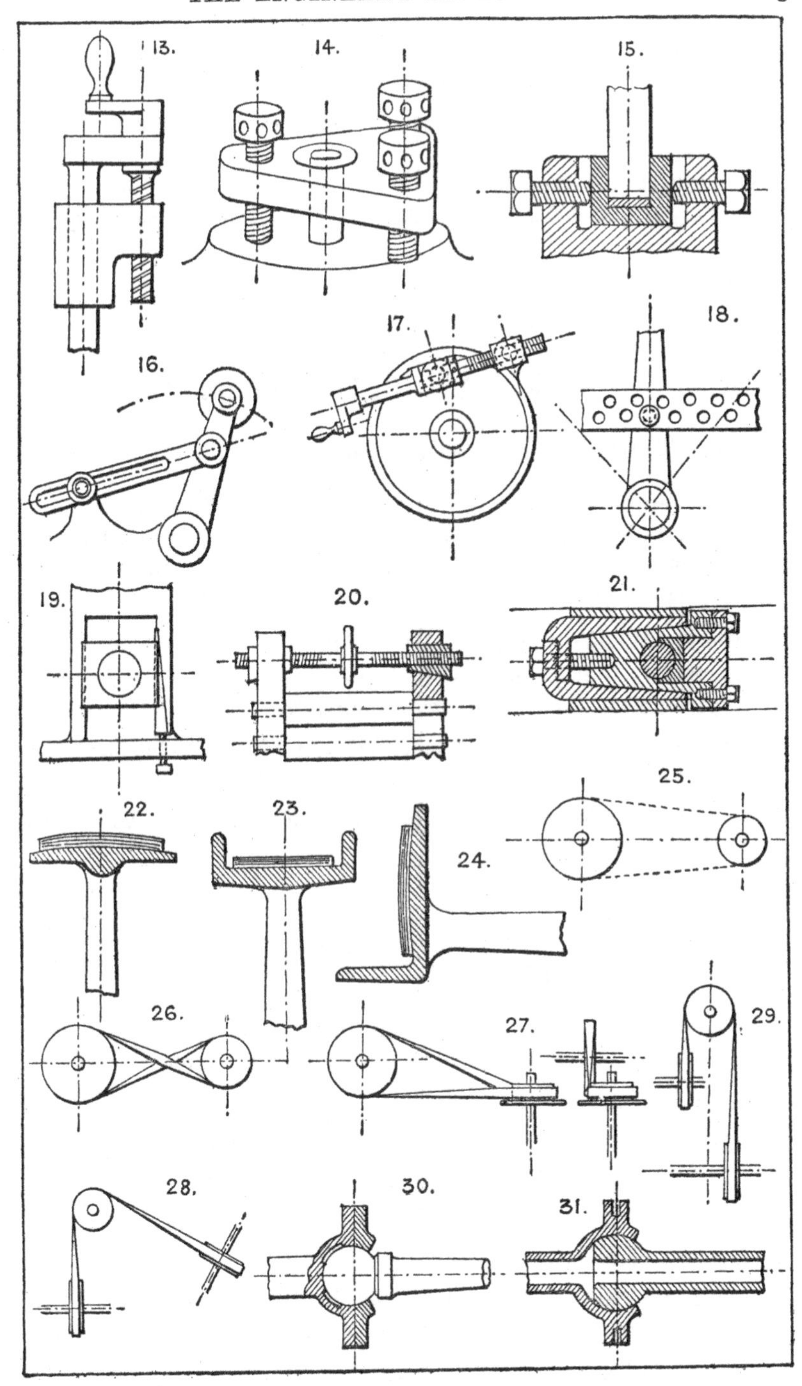
13.
14.
15.
16.
17.
18.
19.
20.
21.
22.
23.
24.
25.
26.
27.
28.
29.
30.
31.

32. Same as No. 16, but with screwed gland. If used without the arm, it forms the ordinary ball castor.

33 & 34. **Dr. Hooke's universal joint.** See application, No. 292. See also No. 1359.

Gas pendants are suspended with a joint similar to No. 31, but the ball, having only a restricted angular motion, is cut down to a segment only.

Section 5.—BRAKES AND RETARDING APPLIANCES.

To retard or arrest motion (revolving or rectilinear).

35. **Strap and lever brake.** The strap is usually faced with wood or leather, but sometimes is used without either. Wood is liable to become noisy. Leather gives the best grip. Iron upon iron, or wood upon iron is not safe if liable to become oily or wet.

36. **Block and lever brake.** Wood or cast-iron blocks are used.

37. **Compound block and lever brake**; avoids putting cross strain on the shaft—used on winding engines, &c.

38. **Internal toggle brake,** employed for friction clutches. See Section 15.

39 & 40. **Double block and lever brake** on wheel rim. The strains are self-contained.

41. **Disc brake**; considerable end pressure is required with this form, and must be arranged for in the bearings of the shaft.

42. **Compound disc brake.** Several discs may be employed, sliding on feathers on the shaft.

43. **Fan brake**; may be run openly in air, or enclosed in a drum with water, oil, or other liquid. (See Allen's patent Governor, &c.)

44. **Spring brake,** for light purposes.

45. **Rope brake or grip,** with toggle motion, and screw for relieving.

46. **Rope brake**: grips by the angular distance of the jaw centres becoming less as the lever end falls.

47. **Rope brake**; with cam lever gripping motion.

48. **Eccentric action lever and block brake.** The eccentric is fixed to the brake lever. This plan also avoids cross strain on the shaft.

49. **Strap and screw brake.**

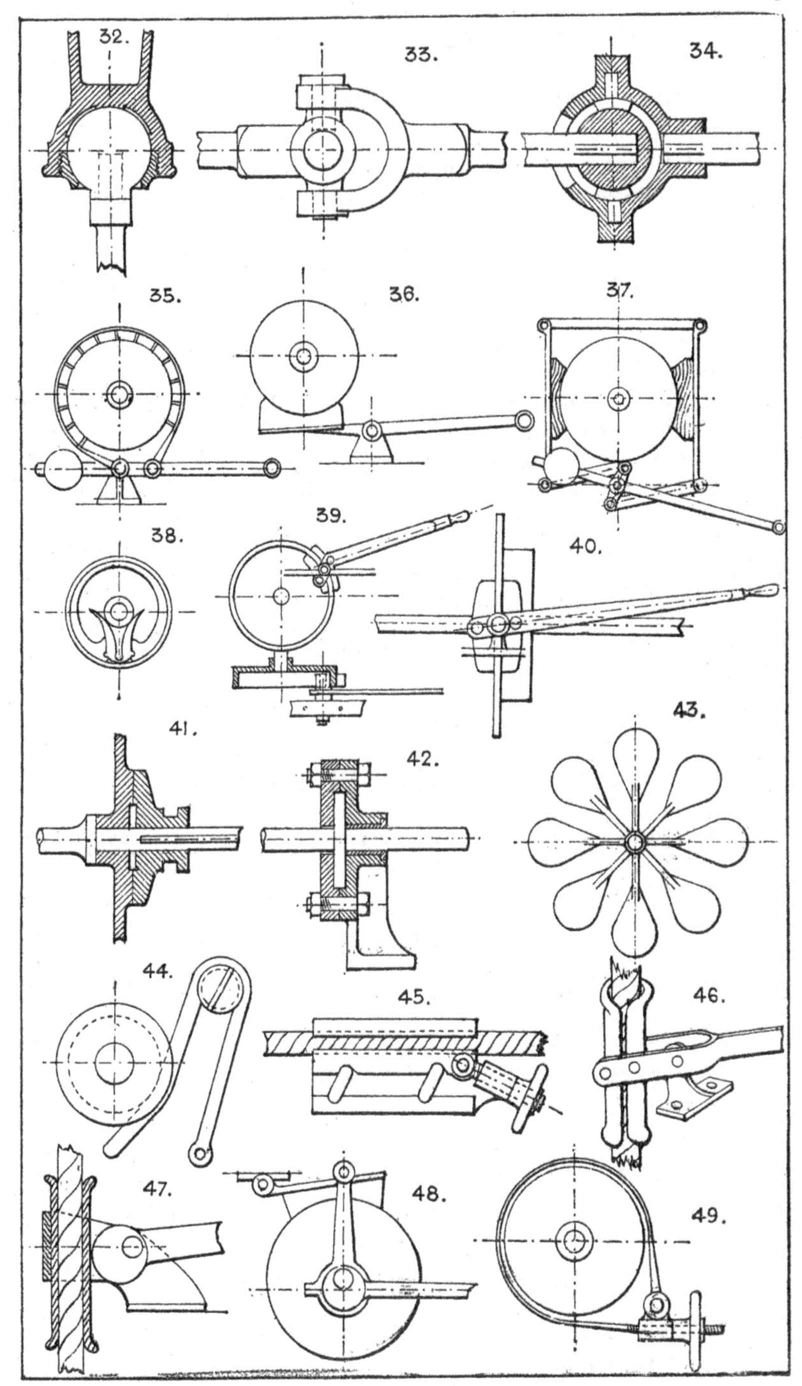
32.
33.
34.
35.
36.
37.
38.
39.
40.
41.
42.
43.
44.
45.
46.
47.
48.
49.

50, 51, & 52. **Three forms of car brakes.** See also the common "skid" or cart brake.

53. **Combined strap and lever brake.** (Fielden's.)

54. **Shaft grip, or brake.**

55. **Centrifugal brake, or clutch.** The weight segments are driven into contact with the ring by centrifugal force. Springs may be used to return them out of action.

56. **Three-segment compound brake:** grips the wheel all round.

57. **Compound bar brake,** with right and left hand screw grip levers, used for heavy gun compressors.

58. **Compound ring brake,** on similar principle to No. 57. See remarks to No. 41.

59. **Wedge and split ring,** used for internal brake ring or clutch, in a similar way to No. 38.

60. **Hollow drums, with radial pockets,** half filled with loose material, or water, mercury, &c., which retard the motion of the drum by the weight and friction of the loose material.

An hydraulic cylinder and piston is frequently used as a brake or retarding device for reciprocating motion, the water passing from one side of the piston to the other, through an adjustable valve. Friction brakes are employed as dynamometers to indicate the power given off or absorbed by any piece of machinery. Automatic brakes (see Sections 15 and 69) are used for hoisting machinery, &c.

Section 6.—TYPES OF BOILERS.

Every conceivable shape of vessel or container has been used as a boiler. Many of the older types are now obsolete, but the following are those most commonly used:—

VERTICAL BOILERS.

61. **Ordinary centre flue boiler.** Sometimes the centre flue is surrounded with tubes, as No. 65.

62. **Vertical multitubular.**

63. **Vertical boiler,** with diagonal tubes and smoke boxes.

64. **Vertical return-flue.**

65. **"Pot" boiler.**

66. **"Field" boiler;** with suspended tubes and internal circulating tubes.

67. **Vertical egg-end boiler,** with spiral flue. Large vertical boilers sometimes have cross flues, or large tubes.

HORIZONTAL BOILERS.

68. **Portable "loco-type" multitubular.**

69. **Fixed return-tube.**

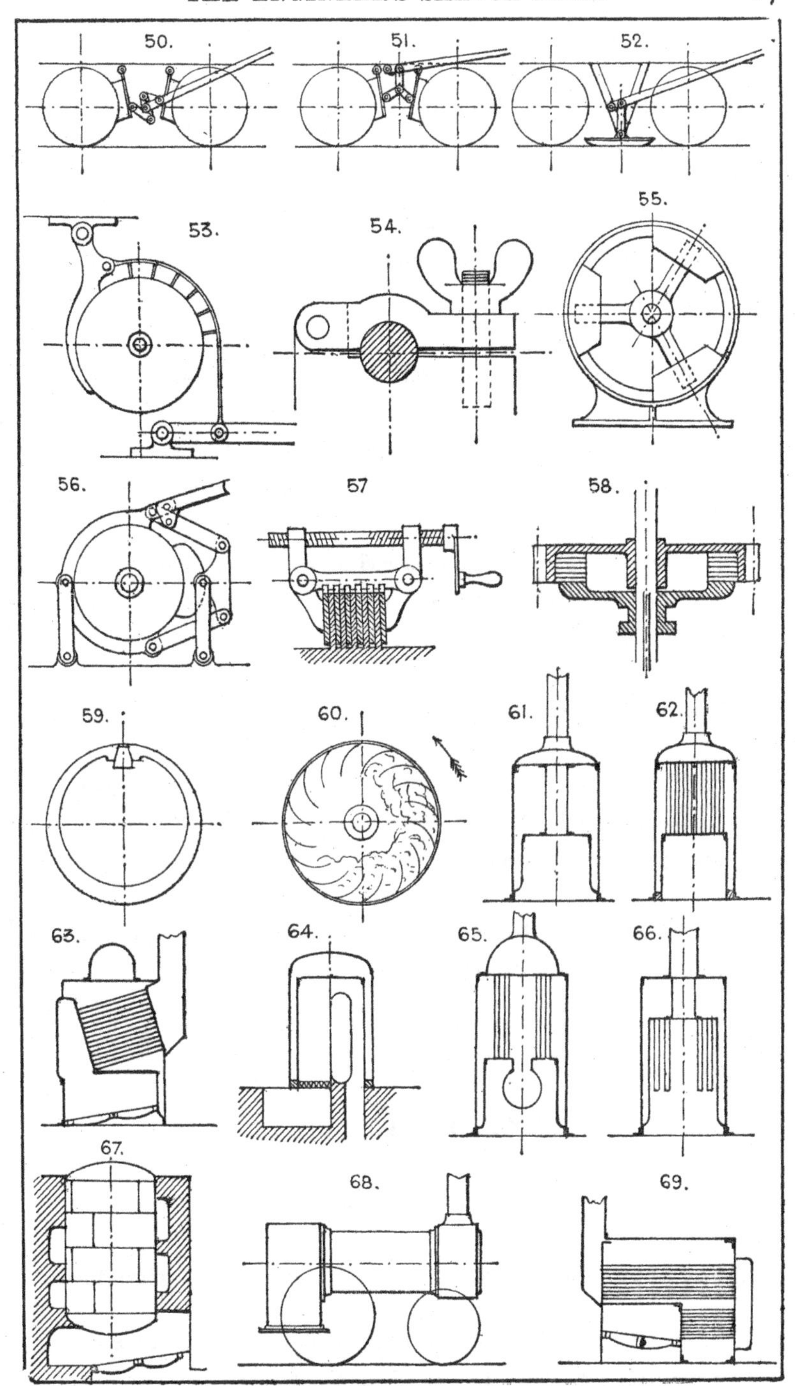
50.
51.
52.
53.
54.
55.
56.
57
58.
59.
60.
61.
62.
63.
64.
65.
66.
67.
68.
69.

70. **Fixed "loco-type"** multitubular; a favorite and useful form, giving good results, and easily cleaned.

71. **Fixed "loco-type,"** with underneath fire-box; sometimes used to economise space, is self-contained, and usually stands on cast-iron feet.

72. **Multitubular-horizontal**; self contained; on cast-iron feet.

73. **Egg-end boiler**; not much used except where the coal burnt per h.p. per hour is not an important consideration.

74. "**Cornish**"; one flue, with enlarged fire-box tube. This type is often made with a parallel flue with cross tubes fixed at intervals throughout its length.

75. "**Lancashire**"; two flues; sometimes has enlarged fire-box tubes.

76. **Oval flue boiler**, with "Galloway" tubes. The Lancashire type is frequently combined with this form by arranging the two circular flues to open into one oval one.

77 & 78. "**Elephant**" **boilers**; employed in connection with coke ovens and other sources of waste heat.

MARINE BOILERS.

79. **Ordinary box form**, with internal fire-box and return flue.

80. **Same type**, but with two fire-boxes and multitubular return tubes.

81. **Underneath fire-boxes** and multitubular return tubes above the fire-boxes, sometimes duplicated, as No. 82.

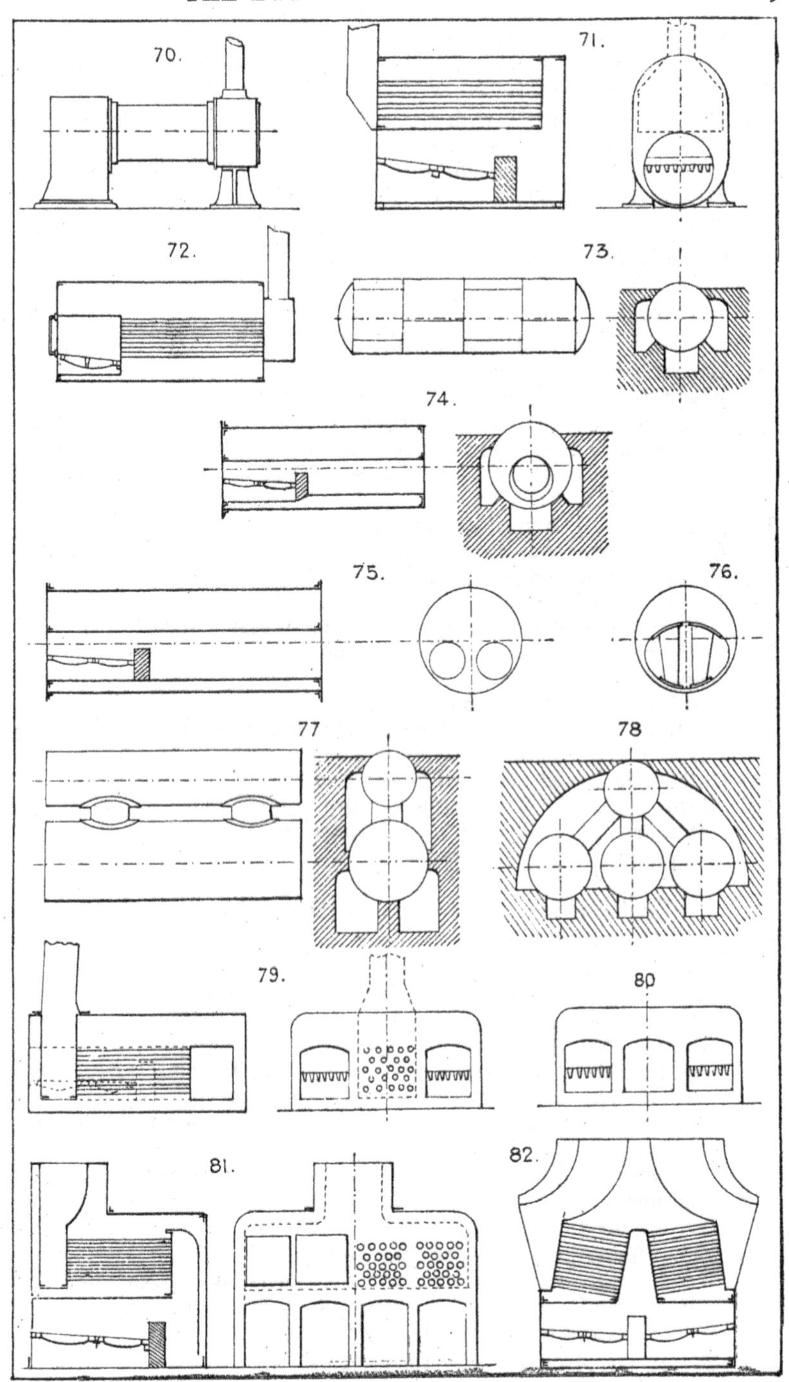
70.
71.
72.
73.
74.
75.
76.
77
78
79.
80
81.
82.

83. **Has two central fire-boxes** and side return-tubes.

The foregoing box patterns are rapidly going out of use, as unsuitable for the higher pressures prevailing with compound engines.

84. **Cylindrical boiler,** with three fire tubes and three sets of return tubes. This form is much used, the surfaces requiring stays being very limited. It is made with double or single fire-boxes.
85. **Cylindrical single flue** and return-tube.
86. **Cylindrical single flue** and multitubular.
87. **Cylindrical double flue** and multitubular.
88. **Cylindrical saddle boiler,** multitubular, used for shallow vessels, launches, &c.

HOUSEHOLD BOILERS.

89. **Kitchen "ell" boiler.**
90. **Kitchen or back boiler,** for ordinary grates
91. **"Saddle" boiler.** The varieties of this type are legion. Every conceivable cross-bridge, water-way, tube, and flue has been added to it by various makers. See Messrs. Graham and Fleming, and other makers' Lists.
92. **Annular cylindrical** greenhouse boiler.
93. **Annular conoidal** greenhouse boiler.
94. **Vertical cylindrical,** closed top greenhouse boiler.

    The last four are types of the greenhouse boilers most in use. They are usually of wrought iron, and all seams welded.
95. **Back boiler** for ordinary register grate.
96. **"Boot" boiler.**
97. **Scullery, or wash-house boiler.**
98. **Scullery, or wash-house boiler, for steam heating.** In public laundries these are usually rectangular in plan.
99. **Coil boiler,** used for small greenhouses, &c.
100. **Sectional, or "Tubulous" boiler.** Root's, and others, are on this principle. They are constructed of simple pipes and T or L pieces usually bolted together.

Section 7.—BLOWING AND EXHAUSTING.

Some of the mechanical blowers are too well known to need illustration here; such are the ordinary Beam Blowing Engine, as in use for blast furnaces, Vertical Blowing Engine, and Horizontal Blowing Engine. In all these a cylinder and piston form the blowing device. Nearly every form of rotary engine (see Section 75) may, by reversal, be converted into a blowing machine. See Root's patent, No. 1307; Baker's, 1325, and others in common use. Fans, centrifugal, (see No. 1337) are still the commonest blowing machines, and are especially suited for light pressures and large volumes of air; but for pressures of from ½ lb. per square inch and upwards, the rotary or cylinder types are best. The following are devices not so well known, but sometimes useful:—

101. **The "Trompe," or water-jet blower.** Water under pressure is discharged through a rose into a funnel-shaped inlet, carrying with it a quantity of air (see Section 45); the water runs off at an overflow, and the air is led away by a pipe.
102. **Steam-jet blower.** (See Section 45.)

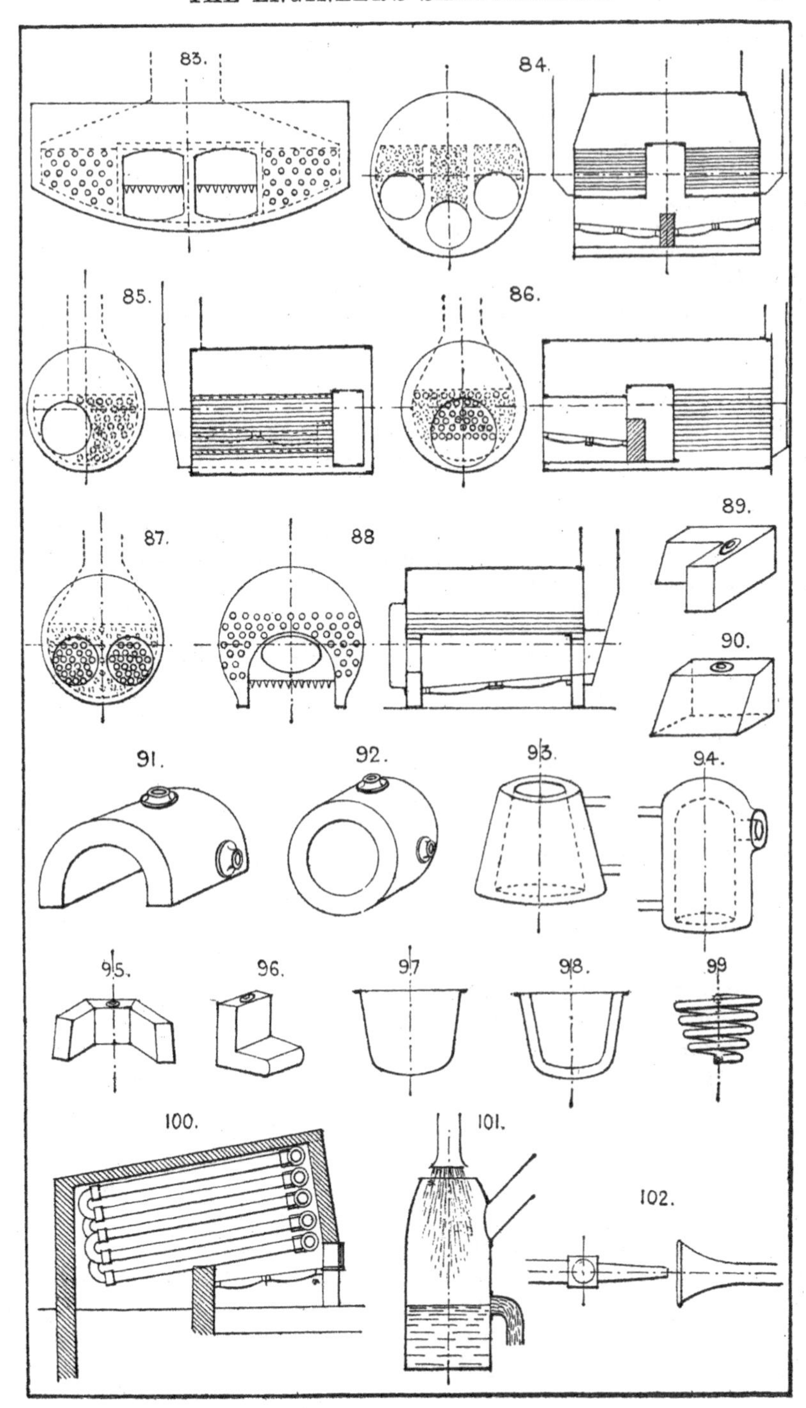
83.
84.
85.
86.
87.
88
89.
90.
91.
92.
93.
94.
95.
96.
97
98.
99
100.
101.
102.

103. **Organ bellows.** The lower "feeders" pump alternately into the double-tier upper "reservoir," which has one set of ribs inverted, as shown, to equalise the pressure throughout its rise. The reservoir is weighted to the required pressure.
104. **Smiths' bellows,** either circular or hinged at one side.
105. **Bell, or gasometer blower,** for light pressures and large volumes.
106. **Regulator, or reservoir,** for blowing engines to steady the blast. The weighted piston serves the same purpose as an air vessel to the ordinary pump.
107. **Disc blower,** with elastic diaphragm piston.
108. **One-crank three-throw blower,** for organs, &c., to give a continuous blast.

Section 8.—BEDPLATES, FOUNDATIONS, AND FRAMING OF MACHINES.

The skeleton framing of a machine for any purpose should be rigid, as light as is consistent with strength and stability (in some cases weight is necessary to minimise vibration), and the ribs, or members of the frame, should be so disposed as to afford the requisite support for all bearings, centres, &c., without redundance; and lastly, symmetry, and a certain degree of elegance and proportion, are desirable. The illustrations are necessarily typical only, and suggestive.

109. **Girder section bedplate** for horizontal distributed bearing, as in a horizontal engine. It may be used double, and the two parts connected by cross pieces and bolts, as No. 112.
110. **Open box bedplate.**
111. **Closed box bedplate.**
112. **Double box bedplate** with cross tie pieces.

    Square or rectangular bedplates are usually of similar sections, stiffened with ribs underneath, and generally cast in one piece.
113. **Side frame and distance rod construction,** suitable for light machines.
114. **Side frames and cross bars** on a base plate. This forms a more rigid construction than No. 113.
115. **Table and legs.**
116. **Rectangular openwork box framing.** Useful for machines with several cross shafts.
117. **Hollow standard** for hammers, vertical engines, and any machine raised above the floor.
118. **Soleplate and standard** for pedestal bracket, &c. Admits of being detached without disturbing the foundation.
119. **Wall box** for shaft bearings, &c.
120. **Arched crosshead for double bearing,** bevil gear, &c.
121. **Wrought-iron sideplate and distance rod construction.**
122. **Wall bracket,** with wall flange, or tongue, to take the vertical strain.
123. **Wrought-iron rectangular bedplate.**

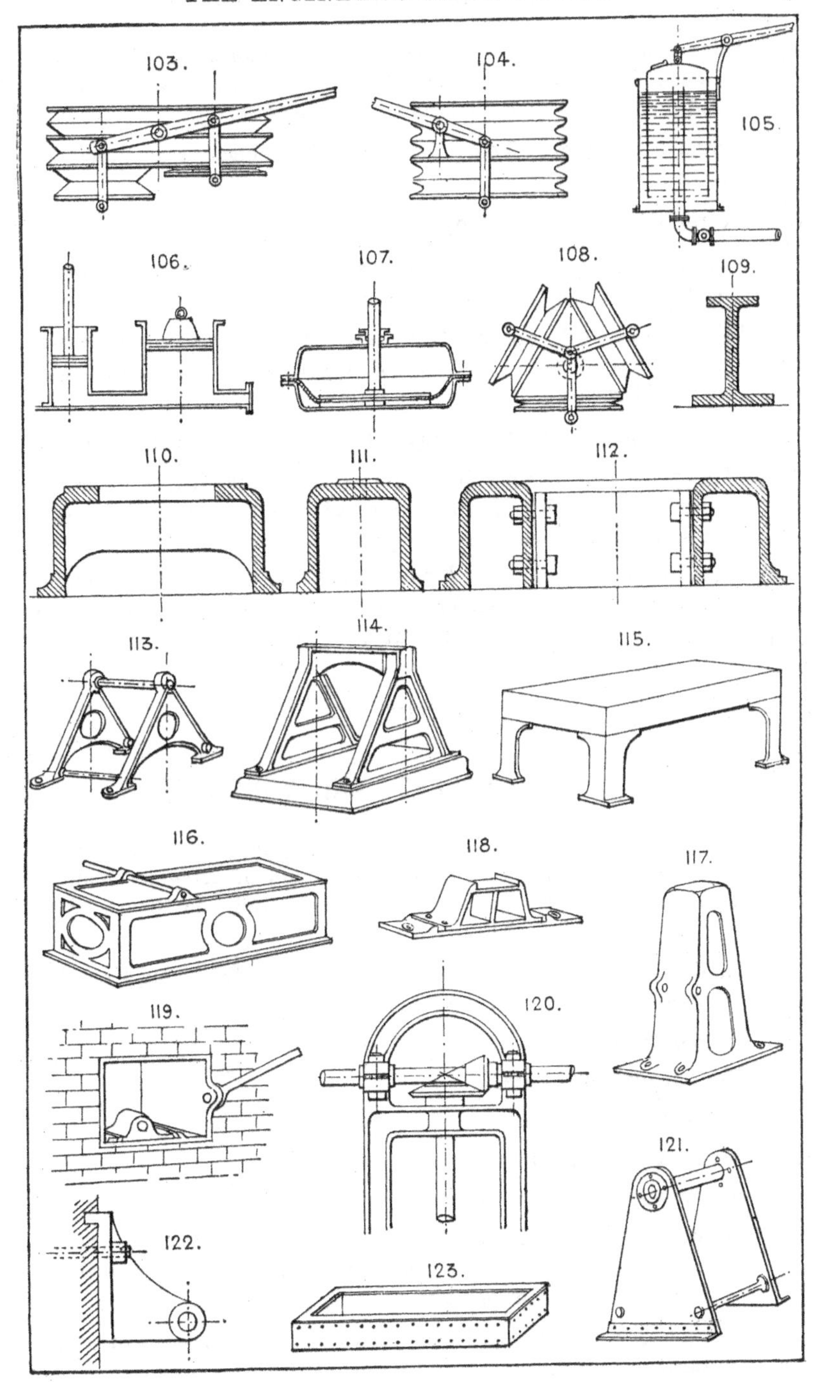
103.
104.
105.
106.
107.
108.
109.
110.
111.
112.
113.
114.
115.
116.
118.
117.
119.
120.
121.
122.
123.

124. **Base plate** for column, &c., with concrete foundation. The bolts are usually T headed (see No. 1404), in open recesses, so as to be easily removed without disturbing the base plate.

125. **Dovetail and key fixing** for brackets, bearings, or any separate detail of framing.

126. **Foundation for box bedplates.**

127. **Vertical columnar,** or distance rod construction, used for marine engines, vertical engines, presses, &c.

128. **Plinth, column, entablature, and cross bracing,** used for beam engines and machinery of a straggling kind with many detached parts.

129. **Flat bar side framing,** strong, light, and cheap, but not very rigid.

130. **Wrought-iron L and flat bar rectangular frame,** suitable where great rigidity is not needed, but where cast-iron is not safe or desirable.

Wrought-iron is becoming much more largely employed for the framing of general machinery than heretofore, and it is customary in many cases to supplement a cast iron base or frame with wrought iron or steel bars.

Section 9.—CAM, TAPPET, AND WIPER GEAR.

For producing, from plain circular, or reciprocating motion, variable speed or motion, also intermittent and every kind of irregular motion. Cams are either open or covered. Nos. 131, 132, and 133 are open cams; Nos. 137 and 138, covered cams.

131, 132, & 133. Three forms of the "**heart**" **cam,** for giving a regular or intermittent vertical motion to a lever end.

134. **Crown cam** for vertical shaft.

135 & 136. **Jumping cams.**

137. **Covered heart cam.**

138. **Covered crown cam.**

139. **Wiper and lever motion.**

140. **Twisted bar** with sliding bush, which travels from end to end of the bar, and being prevented from turning, causes the bar to turn on its axis to the amount of its twist.

141. **Crank pin and slotted lever;** gives a variable speed with quick return.

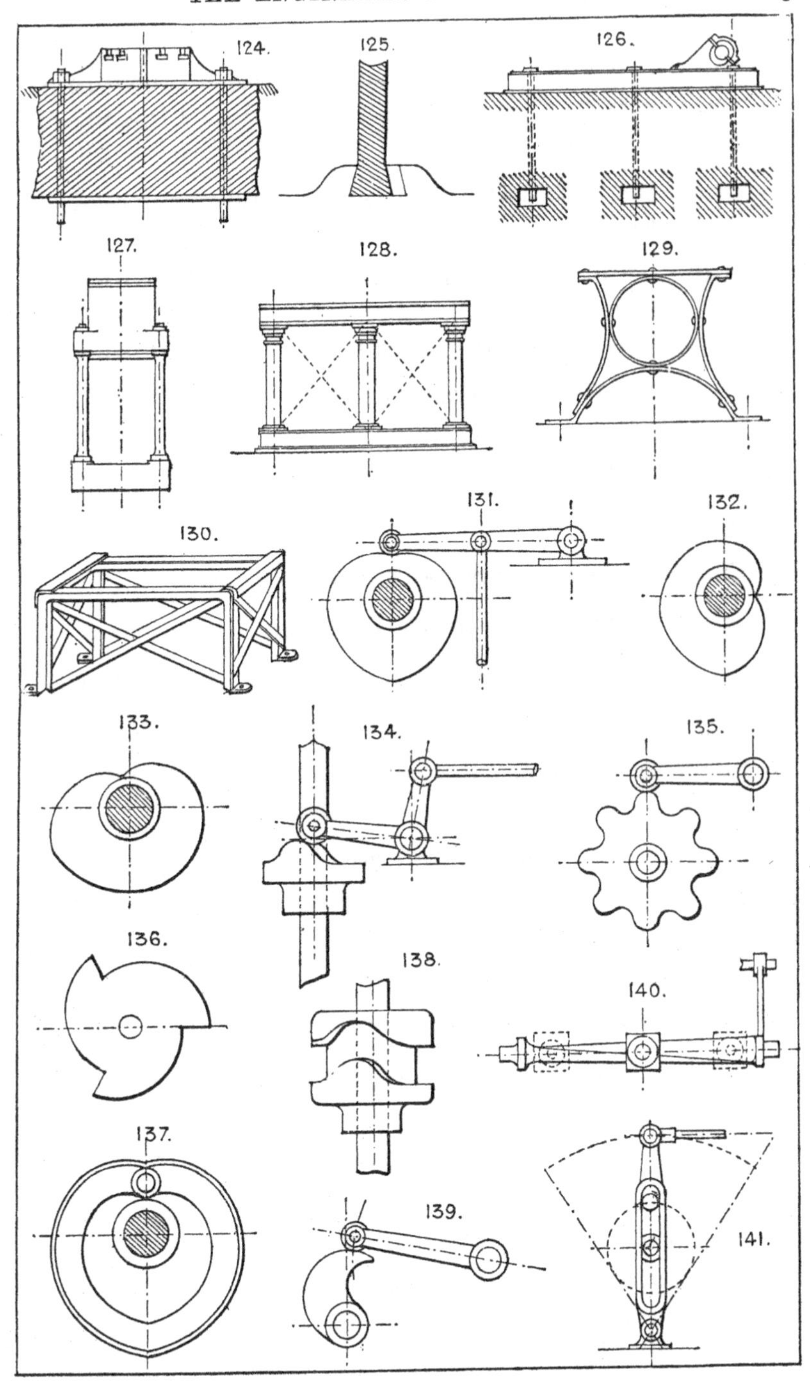
124.
125.
126.
127.
128.
129.
130.
131.
132.
133.
134.
135.
136.
138.
140.
137.
139.
141.

142. **Spiral radius bar** for opening valve. The valve is lifted off its seat by the radial motion of the lever against the inclined radius bar.

143. **Crank pin and slotted lever motion,** with slot arranged for irregular or intermittent motion.

144. **Eccentric and slotted arm.** The pin at the top of the arm has both a vertical and horizontal motion, causing it to trace an ellipse, the pin upon which the slot runs being fixed.

145. **Wiper and lever motion,** with rubbing plate; used for Cornish valves, &c.

146. **Stamp mill.**

147. **Scroll cam.**

148. **Crank and lever,** intermittent or continuous motion.

149. **Piston, or valve rod and lever motion.**

150. **Similar movement,** but with anti-friction roller on end of lever.

151. **Rod and lever reciprocating motion,** with anti-friction roller.

152. **Similar movement,** with a socket forged in the rod and the end of lever rounded to allow for angular motion.

153. **Diagonal disc cam,** or "swash plate."

154. **Motion for belt shifting** with dead travel at half stroke. This allows the lever to move a certain distance on each side of the centre without moving the belt shifting bar.

155 & 156. **Sectors and bent lever,** used on Cornish engine valve gear.

157. **T lever valve motion,** used in rock drills, some forms of steam engines, &c.

158. **Four-bolt camplate,** used for screwing dies, locks for fireproof safes, &c.

159. **Slot, cam, and lever motion.**

160. **Barrel motion** for musical instruments, looms, &c., in which the barrel is provided with pins or staples to lift the respective levers.

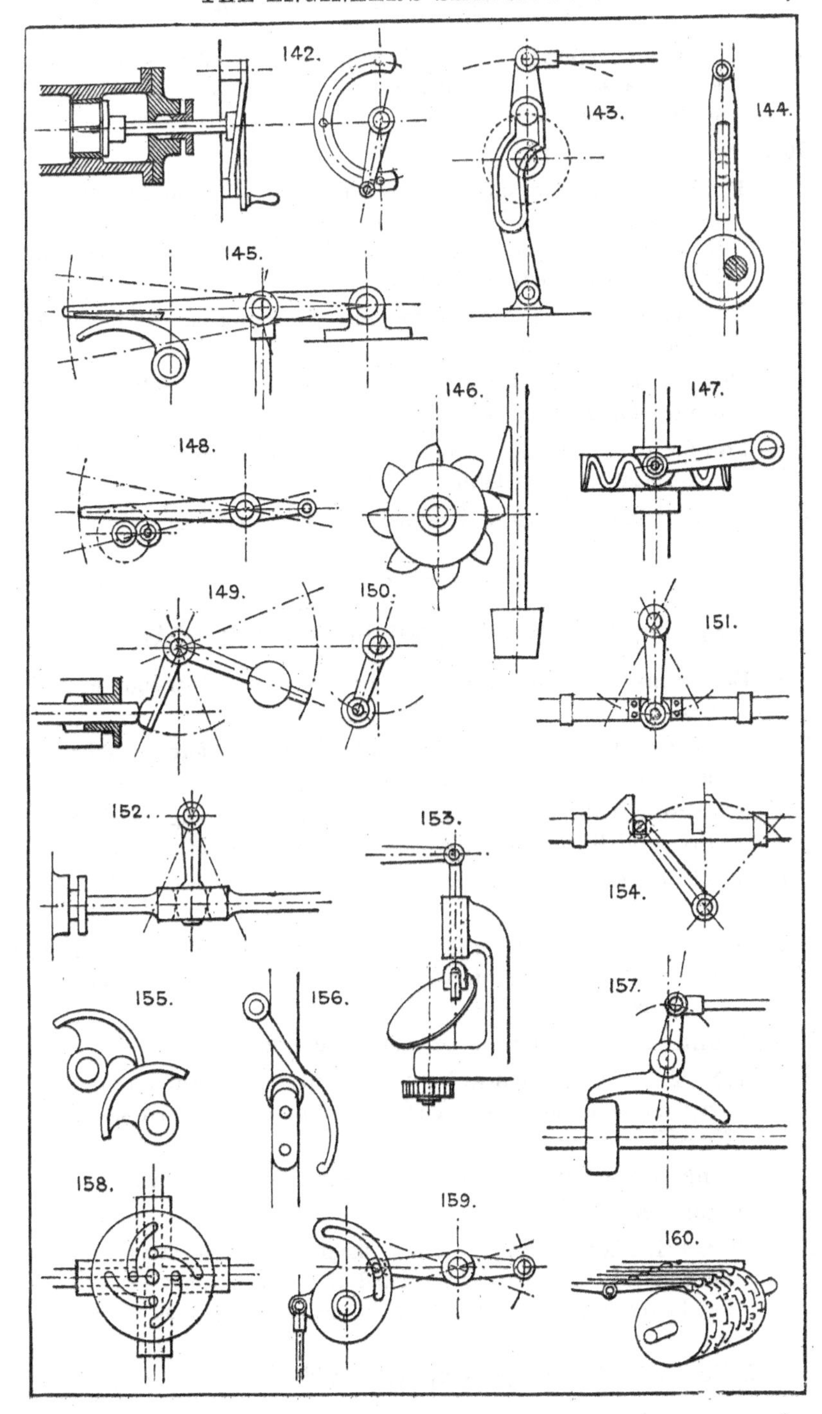
142.
143.
144.
145.
146.
147.
148.
149.
150.
151.
152.
153.
154.
155.
156.
157.
158.
159.
160.

161. **Drum with spiral vanes** of long pitch, operated by a revolving arm on a shaft at right angles to the cam shaft, used for intermittent circular motion.

162. **Volute and lever.**

163. **Double screw,** for converting circular into reciprocating motion; has a right and left hand screw thread, and a shuttle attached to the lever end shaped to fit the thread, and capable of swivelling to the angle of the screw thread.

164. **Eccentric ring and roller motion,** for converting circular into reciprocating motion.

165. **Triangular cam.** Gives three reciprocations to the sliding bar in one revolution of the cam.

166. **Fan** for giving motion to several rods or arms at one time, used for organ composition pedal movement, &c.

167. **Crossed lever motion** with inclined contact surfaces.

Section 10.—CRANK AND ECCENTRIC GEAR.

168. **Bent crank** of round section; retains the fibre and strength of the metal.

169. **Square forged crank.** The crank arm is usually forged solid and the slot cut out by machine.

170. **Built-up crank.** There are other methods of building them. See *Mechanical World*, December 1885. See also No. 182.

171. **Single crank,** usually of wrought iron, but often made with cast-iron arm.

172. **Disc crank.** This form is generally adopted when cast iron is employed, and the counterbalance weight cast upon it, to balance the connecting rod, &c.

173. **Counterbalanced single or double crank.**

174 & 175. **Two forms of crank pin eccentrics;** sometimes used to drive the slide valve instead of the ordinary sheave and strap, as No. 183.

176. **Crank pin set in a boss formed on the driving wheel.**

177. **Double rod crank.**

178, 179, & 181. **Hand cranks.** These should always be fitted with a loose ferrule of wood for the hand if possible, as much power is lost by the slipping of the hand to change its grip as the crank revolves.

180. **Solid three-throw crank shaft,** turned out of a solid forging.

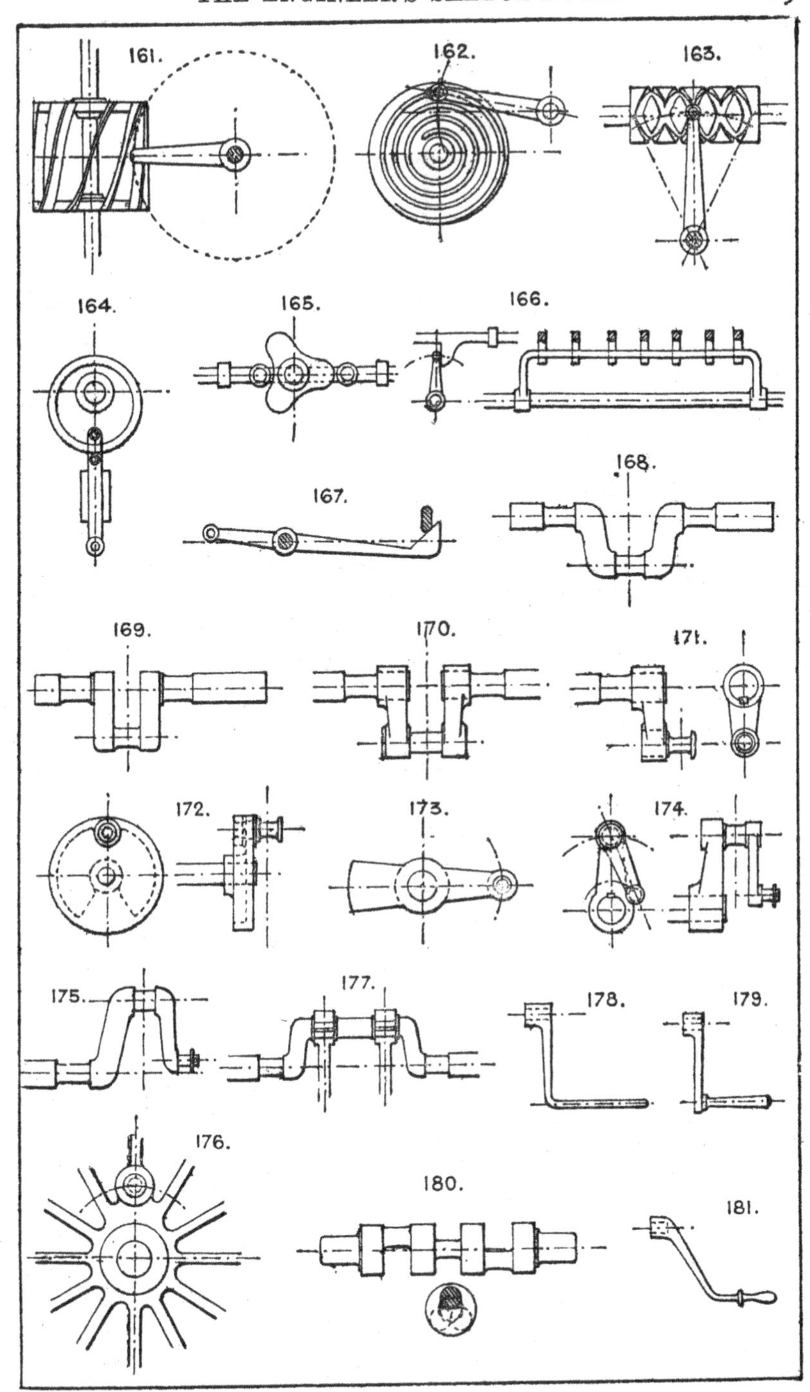
161.
162.
163.
164.
165.
166.
167.
168.
169.
170.
171.
172.
173.
174.
175.
176.
177.
178.
179.
180.
181.

182. **Built crank.** Several modifications of this are in use for large marine shafts.

183. **Solid sheave eccentric.**

184 & 185. **Split sheave eccentrics.**

Large eccentrics cause great loss from friction, unless provided with friction rollers in the sheave; but are sometimes used to avoid an additional crank in shaft.

186. **Eccentric motion** for multiplying travel of eccentric by leverage.

187. **Crank motion to turn an angle** instead of bevil gear and shaft, the cranks being of the form of Nos. 174 or 175.

188. **Shifting or variable throw eccentric.** The sheave is slotted to fit the shaft, and its throw is governed by a disc having a spiral slot and locking bolt.

189. **Another form of shifting eccentric.** The sliding block is arranged to lock in any part of the slot in the sheave.

190. **Another form of shifting eccentric,** in which the sheave is loose on an eccentric boss cast with the worm wheel, and is revolved by the worm, the bearings of which are fixed to the sheave.

See also Nos. 606, 712, 720, 728, 729.

See also Sections 40 and 79.

Section 11.—CHAINS AND LINKS.

For Hooks, Swivels, &c., see Section 43.

191. **Ordinary long or short link chain.** It is sometimes made to exact pitch to fit a snug or sprocket wheel. See Nos. 1250 & 1251.

192. **Stud link chain.**

193. **Flat chain** for use on flat rim double-flanged pulley.

194. **Square link pitched chain** for sprocket wheel.

195. **Stamped link pitched chain** and special sprocket wheel.

196. **Ordinary pitched link chain.** Links drilled to templet.

197 & 198. **Pitched chains** to drive wheels with ordinary and special teeth.

199. **Another form of square link chain.**

200. **Stamped link chain,** for light purposes.

201 & 202. **Long link flat suspension chains.**

203. **Gib and cotter attachment** for long link flat suspension chains.

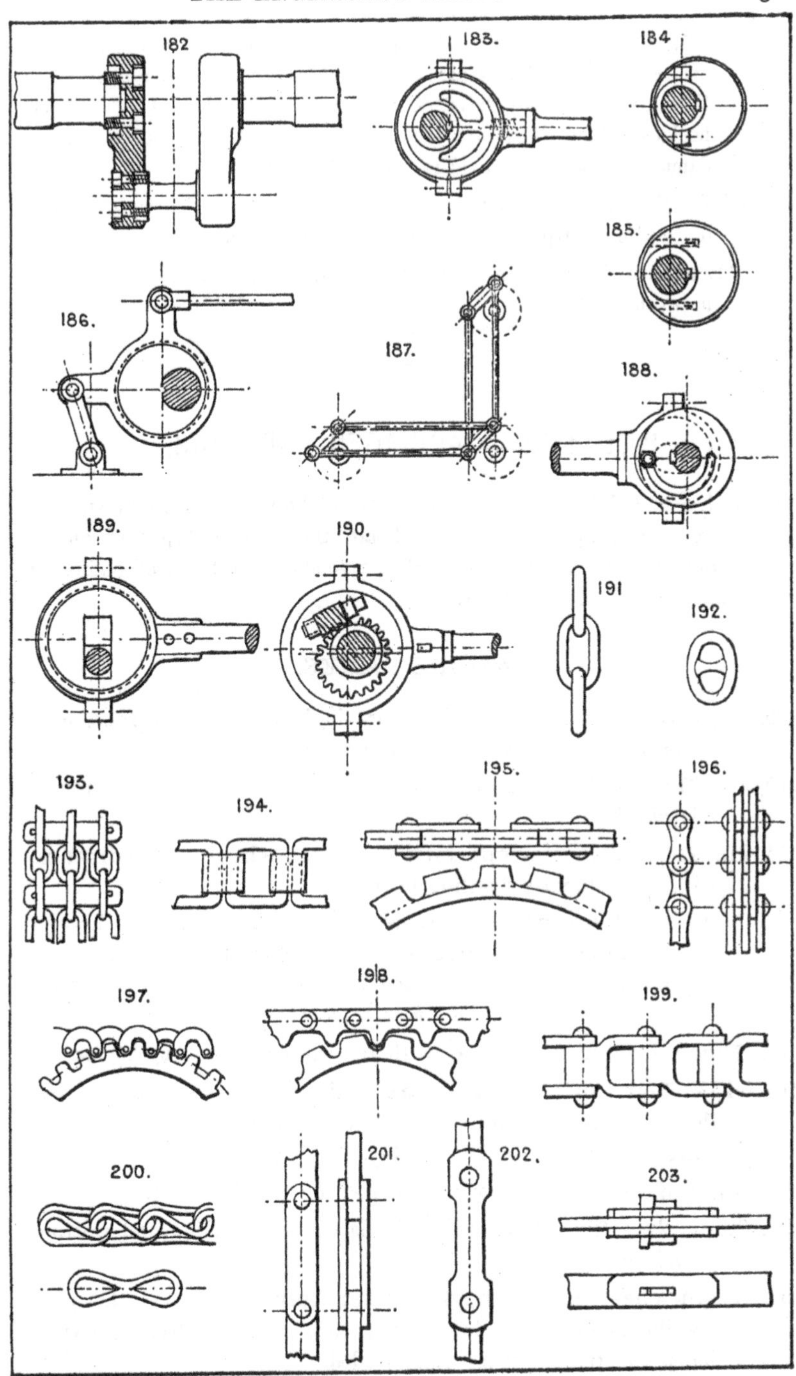
182
183.
184
185.
186.
187.
188.
189.
190.
191
192.
193.
194.
195.
196.
197.
198.
199.
200.
201.
202.
203.

204 & 205. **Drive chains.** See Ewart's patent, No. 2752–76, and others. These chains are replacing belts for many purposes, as they give a positive drive, do not stretch so much, and last longer, besides which they are easily detached at any point, and a damaged link can be readily replaced.

206 & 207. **Thrust chains,** with friction rollers at each junction, used in hydraulic multiplying cylinder gear in some cases.

208. **Ewart and Dodge's patent chain,** with renewable seatings between the links.

Section 12.—CARRIAGES AND CARS.

The design and details of these must always be suited to circumstances. We only propose here to indicate the various types of under-framing and wheels in use, and to give sketch sections of bodies or cars for different purposes.

UNDER-FRAMES.

209. **Two-wheel suspension car** for single rail or wire rope, used commonly on some kinds of cranes. See Section 18.

210, 211, 212, & 213. **Three-wheel cars.** See also the various types of tricycles in use.

214, 215, 216, & 217. **Various forms of four-wheel under-frames,** with and without swivelling bogies.

A car with four wheels arranged as No. 217, but with the leading and trailing wheels slightly raised off the ground, is used as a goods car or hand truck, and is very readily swivelled about, running, of course, actually on three wheels only.

218. **Five-wheel under-frame,** with and without swivelling bogies.

219 & 223. **Plans of six-wheel cars,** with swivelling gear for curves; the centre pair having end play, swivel the leading and trailing axles by means of the jointed stays.

220. **Plan of four-wheel car,** with swivelling gear for curves.

221, 222, & 224. **Six-wheel cars,** the latter with leading and trailing swivelling bogies.

225. **Eight-wheel double-bogie under-frame.** This is the plan usually employed in long cars; each bogie is free to swivel independently, and is centrally loaded.

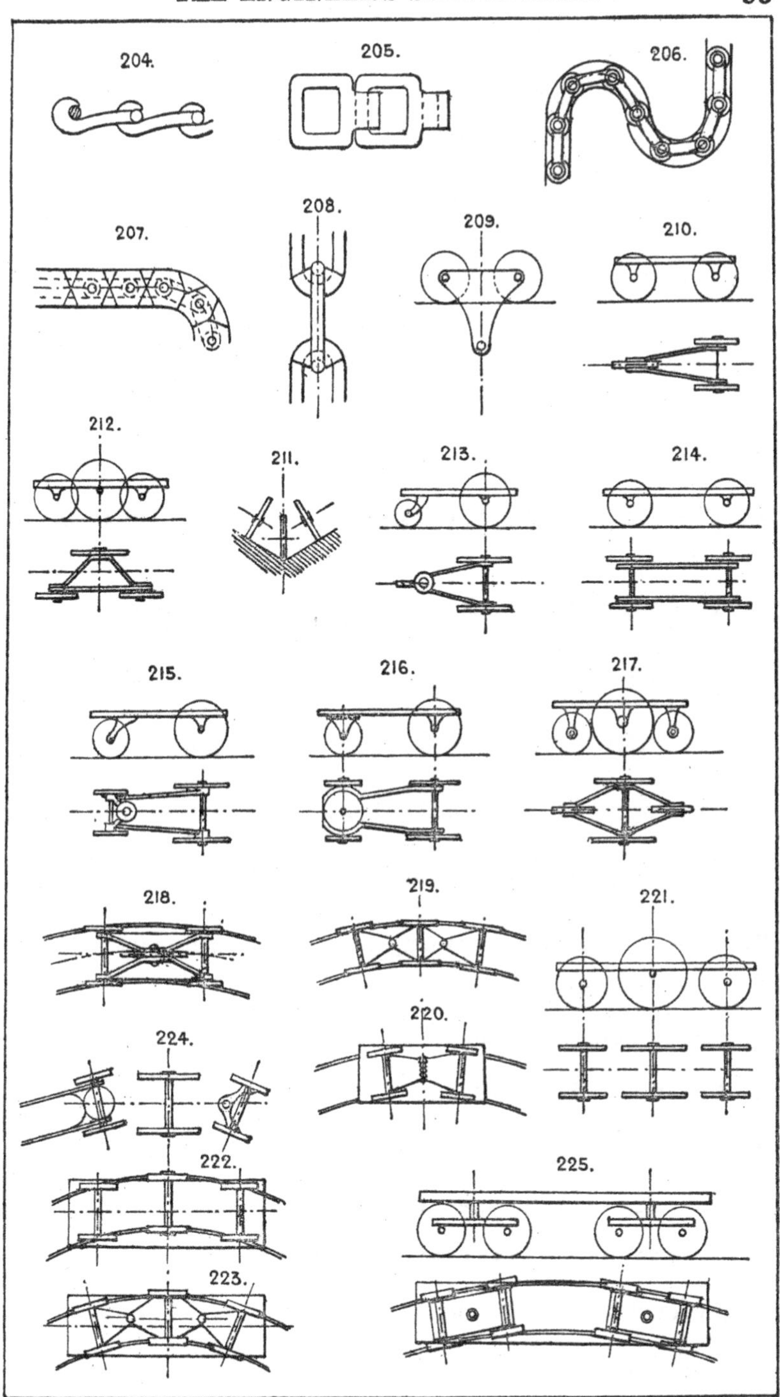
204.
205.
206.
207.
208.
209.
210.
212.
211.
213.
214.
215.
216.
217.
218.
219.
221.
220.
224.
222.
225.
223.

226. **Ten-wheel double-bogie under-frame,** the centre pair to have end play or broad flat tyres.

227. **Twelve wheels and three bogies.** The centre bogie must have end play, either as in Nos. 222 or 226, or with transverse rollers between the bogie and frame.

Note that in Nos. 221, 223, & 224 the centre pairs, if running on rails, must have either end play in the bearings or flat broad tyres.

228. **Open passenger car,** either with transverse or longitudinal seats.

229. **Covered passenger car,** with either longitudinal or transverse seats.

230. **Passenger car,** with outside and central longitudinal seats.

231. **Passenger car,** with upper and lower longitudinal seats.

232. **Passenger car,** as No. 231, but with seats reversed.

233. **Passenger car,** for one-rail railway.

234. **Passenger car,** similar to No. 230, but with seats reversed.

235. **American plan of passenger car,** with transverse seats and central gangway.

236. **Goods cars,** low sided.

237. **Covered or box wagon.**

238. **Hopper wagon** for discharging below.

239. **Side discharge hopper wagon.**

240. **Side tip** (or end tip) **three-centre wagon.**

241. **Tip cart.**

242. **Tip wagon.**

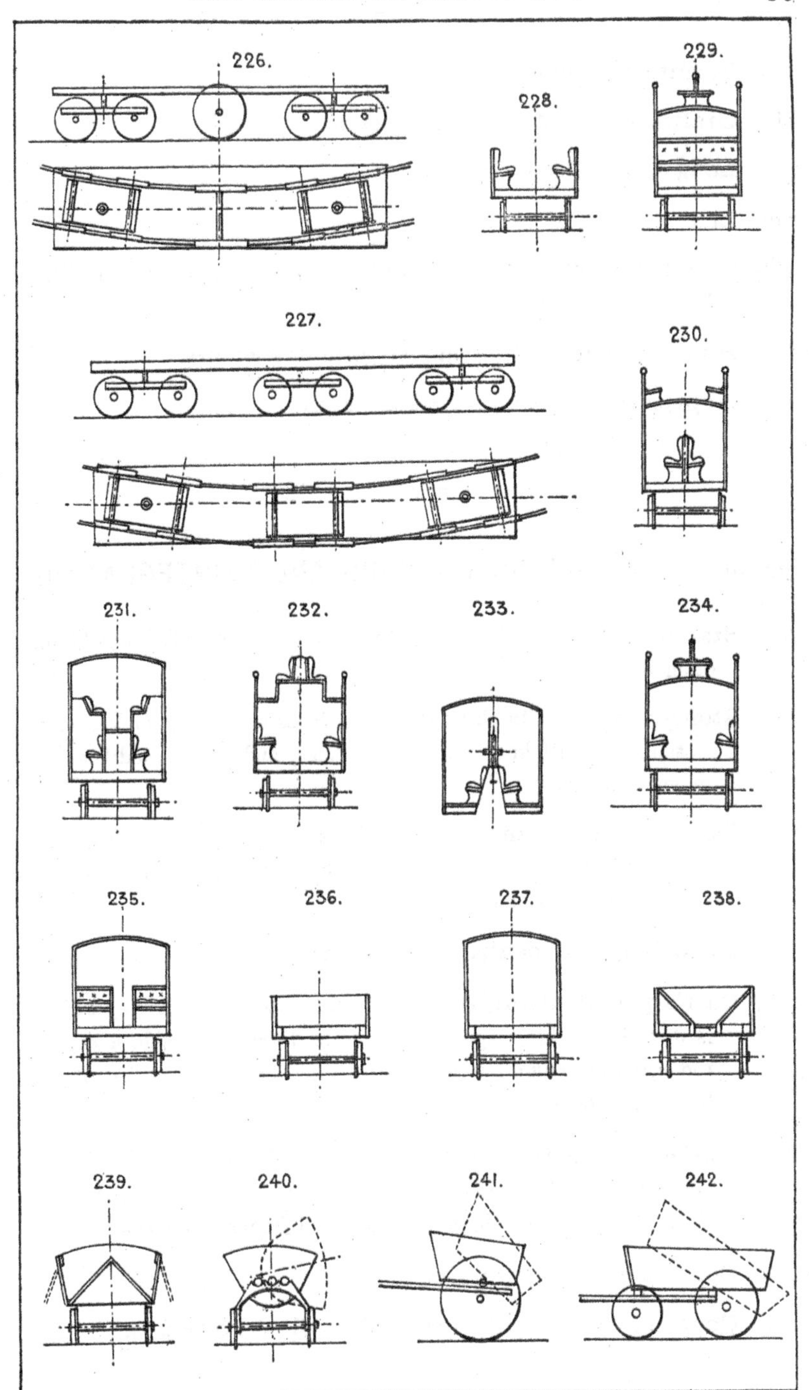
226.
227.
228.
229.
230.
231.
232.
233.
234.
235.
236.
237.
238.
239.
240.
241.
242.

243. **Furniture wagon.**

244. **Grafton's patent side tip wagon.**

245. **Long truck for boilers, &c.**

246. **Incline car for passengers.**

247. **Segmental swivelling bearings,** used instead of a swivelling bogie and centrepin.

248. **Hudson's patent tip wagons,** with three centres.

249. **Hopper wagon,** with central discharge.

Section 13.—CRUSHING, GRINDING, AND DISINTEGRATING.

250. **Stamp mill,** generally arranged in a battery of 4 or 6, for gold and other ores.

251. **Stone-breaker,** with chilled iron jaw faces and toggle or knapping motion. See Blake's, H. R. Marsden's, and other modifications in common use.

252. **Double edge-runners.** Sometimes driven below. In some designs the rollers revolve, and in others the pan revolves and the roller shaft is stationary.

253. **Lucop's patent centrifugal pulveriser.**

254. **Carr's patent disintegrator.** In this machine, each ring of bars is driven at a high speed in opposite directions inside a casing, the material being broken by the rapidity and intensity of the blows it receives.

255. **Horizontal centrifugal roller mill.** The material is crushed between the rollers and the shrouding of the pan by the centrifugal force of the rollers, which are suspended from a crosshead.

256. **Cone roller mill,** with vertical spindle.

257. **Cone roller mill,** with horizontal spindle and conical pan.

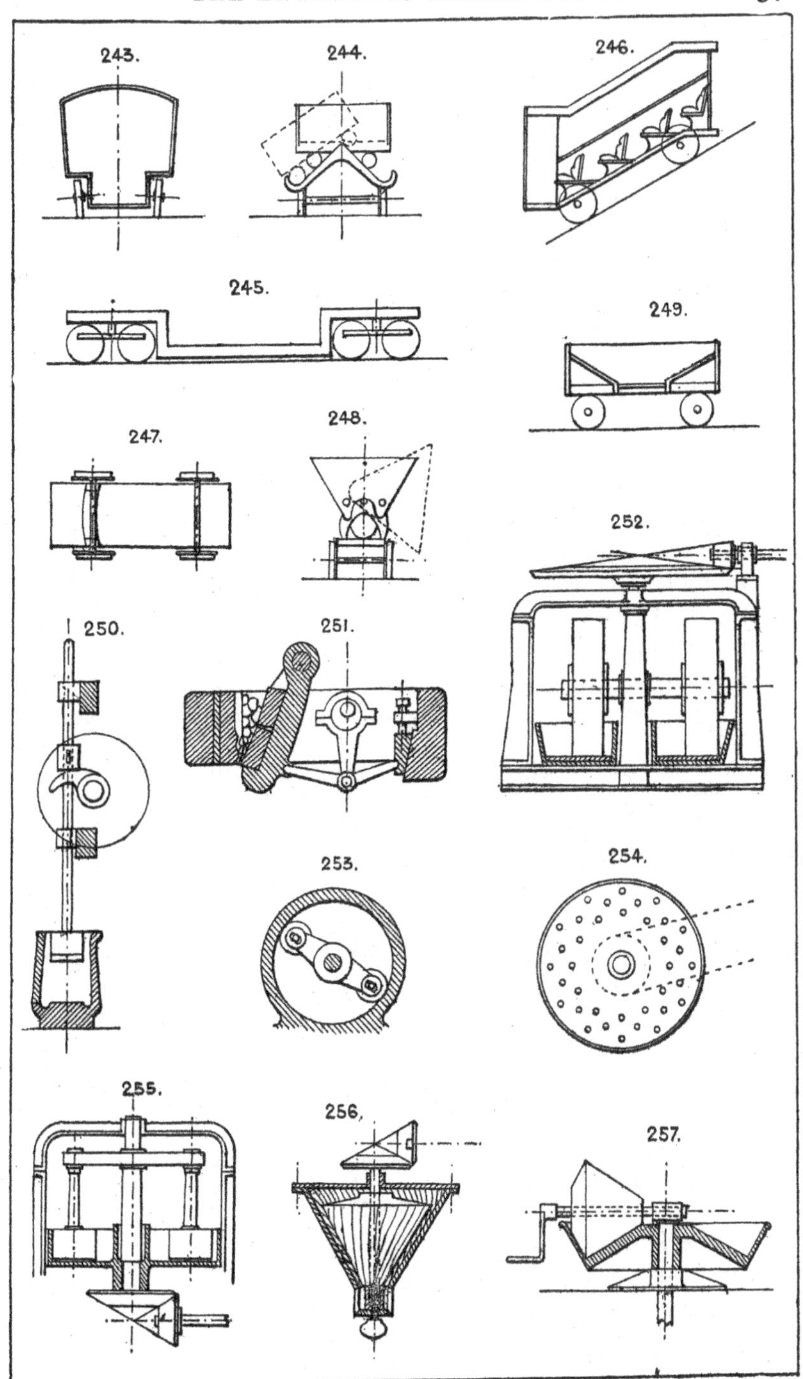
243.
244.
246.
245.
249.
247.
248.
252.
250.
251.
253.
254.
255.
256.
257.

258. **Enclosed cone roller mill,** with horizontal spindle and spirally grooved roller and casing.

259. **Toothed sector mill.**

260. **Conical edge runner and pan.**

261. **Ordinary flour mill.** The material is fed in the centre, passes between the stones, and falls out into the outer casing.

262. **Rattle barrel,** for cleaning and burnishing articles by mutual attrition; sand or emery is sometimes used to assist the process.

263. **Ball and pan mill,** for crushing ores, &c. The balls are carried round by a cross arm fixed to the central spindle.

264. **Inclined ball and pan mill.**

265. **Oscillating mill.**

266. **Drum and roller revolving mill.**

267. **Cradle and roller mill.**

268. **Cone disc mill;** the cones being inclined axially to one another, the material is crushed at the lower side of the cones.

270. **Horizontal cone plate mill.**

271. **Revolving stamp and pan mill** for ores.

272. **Vertical cone mill.**

273. **Revolving pan and ball mill.**

274. **Planishing discs** for accurately rounding iron bars. See the patent rolled shafting in use, manufactured by the Kirkstall Forge Co. and others.

275. **Vertical cone grinding and crushing mill.** The vertical shaft has an eccentric motion at the footstep, giving a swaying rotatory motion to the grinding cone.

276. **Crushing rollers** with spring bearing.

Section 14.—CENTRIFUGAL FORCE, APPLICATIONS OF.

277. **Centrifugal drill.** The cross bar A is alternately pressed down and allowed to rise, the strings winding on the spindle alternately in opposite directions.

a. **Fly-wheel.** Use; to receive and store redundant motive power, and give it off again when the motive power falls below the average.

b. **Centrifugal hammer.** One or more hammers are loosely jointed to a revolving boss, and strike rapid blows on an anvil fixed in the path of their circumference. See No. 1915.

c. **Pulverising machines.** See Nos. 253, 254, 255.

d. **Speed governors.** See Section 41.

e. **Cream skimmers** have a pan revolving horizontally in which the new milk is poured. The cream travels to the outer edge and runs over into a receiving trough.

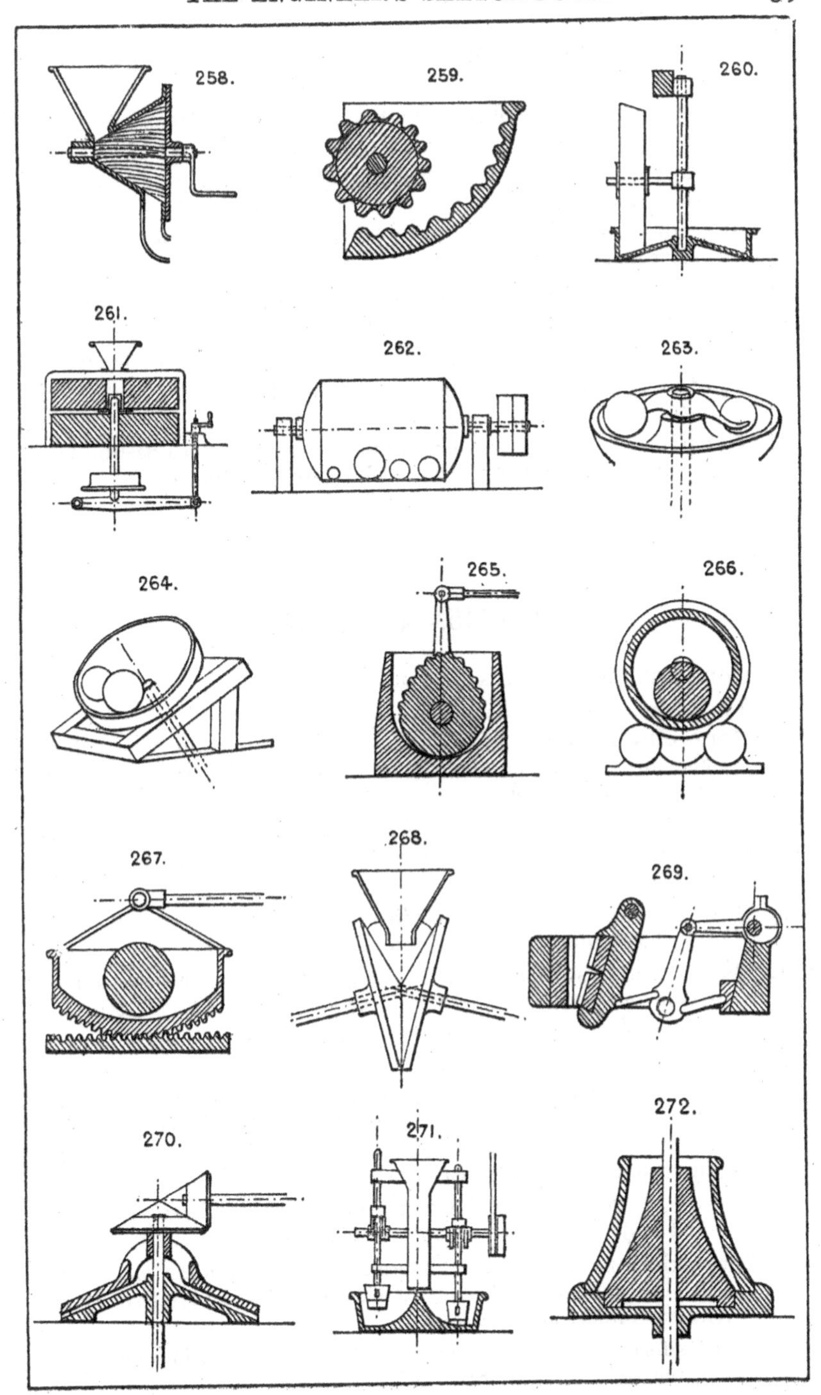
258.
259.
260.
261.
262.
263.
264.
265.
266.
267.
268.
269.
270.
271.
272.

f. **Centrifugal dryers.** Manlove and Alliott's, also Robinson's continuous feed ditto, are examples.

g. **Some forms of turbine.** On the principle of Hero's Eilipile, No. 1696.

h. **Swings. Roundabouts.** Various toys. The *Gyroscope* and tops constructed on its principle.

i. **Juggling and other tricks** performed with pivoted plates and other common articles.

j. **Rattle barrel, or revolving drum,** for polishing small castings, &c., by centrifugal motion and mutual friction, similar to No. 262.

k. **Various machines for grading wheat, grain, and seeds.**

l. **Centrifugal filter** for sugar; a modification of the centrifugal drying machine. See No. 475.

m. **Centrifugal pumps** are forms of fans or turbines (see Section 90); Gwynne's, Schiele's, Andrews', and others are examples.

Section 15.—CLUTCHES.

278. **Common jaw clutch** sliding on a feather key.

279. Two forms of jaws for ditto.

280. **Cone clutch.** Screw gear should be used to operate this, as it is liable to "seize," and there is considerable end pressure on the shaft to be allowed for.

281. **Face (friction clutch)** with V grooves. See remarks to No. 280.

282. **Friction clutch** with three or more segments. See also Nos. 38 and 59.

283. **Pin and hole clutch.** The pin and holes can of course be made to act horizontally instead of radially.

284. **Cam clutch,** used for dexter treadles, also for reciprocating motions driving one way and running loose the opposite way. See also Section 62, Nos. 1135, 1178, &c.

285. **Crank pin and arm driver.**

286. **Pickering's self-sustaining clutch** for hoists. The box A only is keyed to the shaft, and drives the chain wheel and sleeve B by jamming it with the flange of the ratchet-wheel sleeve C by the sliding action of the toothed faces at D.

Several other forms of this clutch are in use. Edwards', Stevens and Major's, and others may be consulted.

287. **Disc friction clutch,** with intermediate leather discs and screw clamping appliance. Mather and Platt's and Addyman's patent friction clutches are examples.

Numerous forms of friction clutches are in use, modifications chiefly of Nos. 38, 59, and 282. See also Section 5.

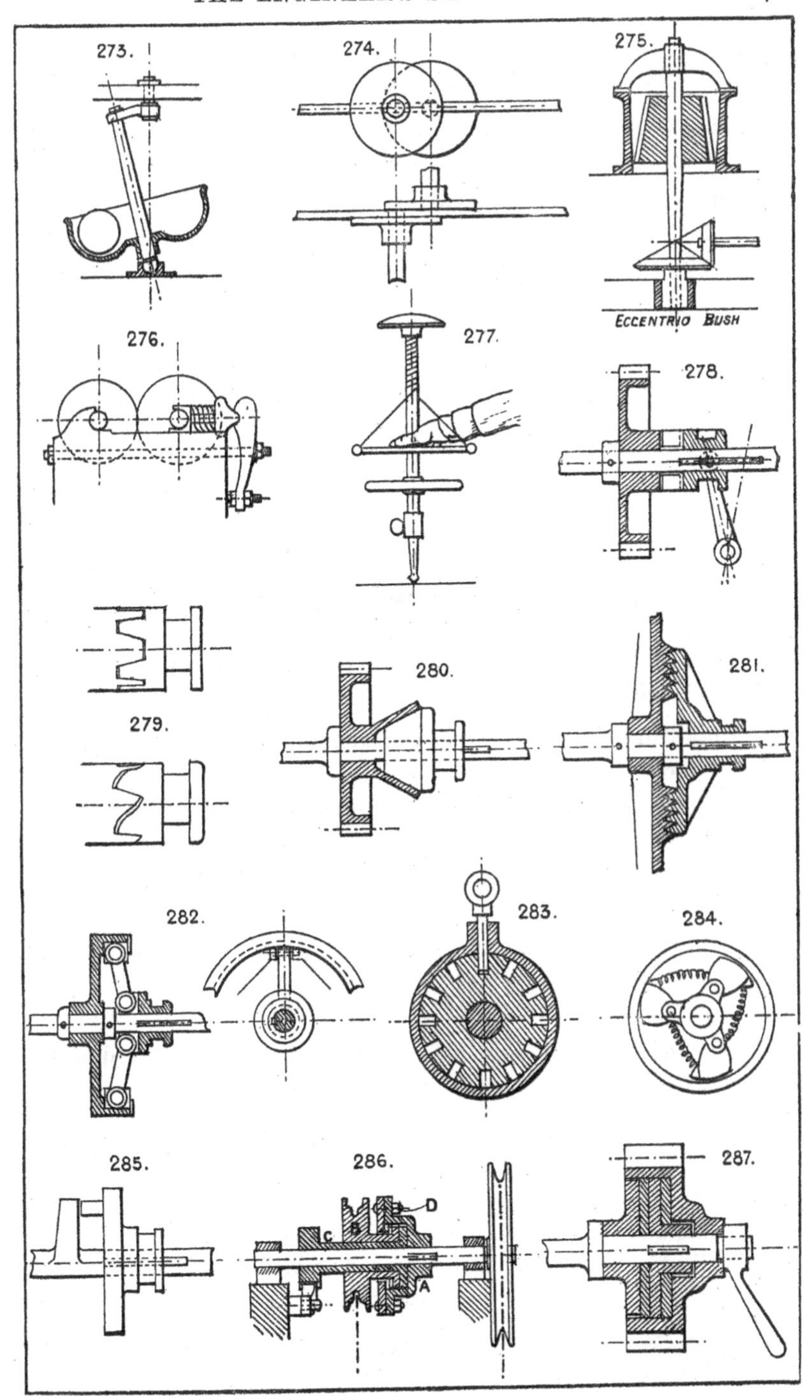
273.
274.
275.
ECCENTRIC BUSH
276.
277.
278.
279.
280.
281.
282.
283.
284.
285.
286.
D
B
C
A
287.

Section 16.—COUPLINGS FOR SHAFTING.

288. **Ordinary flanged coupling,** usually made so that the end of the shaft forms a spigot joint with the opposite half of clutch.

289, 290, 291. **Sleeve couplings.** See also No. 1430 and Section 57. Butler's patent frictional coupling, Kirkstall Forge Co., Leeds, Seller's double-cone vice coupling, and others, are sleeve couplings.

292. **Angle coupling** on Dr. Hooke's principle.
293. **Flexible angle coupling** for light work.
} See also Nos. 33, 34, and 732.

294. **Flanged coupling,** with cross feather or key. This plan gives great torsional strength, especially if the coupling flanges are forged solid with the shafts.

Section 17.—CONNECTING RODS AND LINKS.

295. **Turned and finished link** without any adjustments; ends may be solid, or forked as No. 297.

296. **Flat link** of similar description, with raised bosses for facing and wear.

297. **Adjustable link,** with right and left hand screw coupling. Lock nuts may be added to prevent the coupling working back.

298. **Strap link,** fitted with brasses, gibs and cotters, and distance bar. In this link the wear of brasses is all taken up one way by the gib and cotter; therefore, if great accuracy in the distance apart of centres is necessary, gibs and cotters should be fitted at both sides of one pair of brasses, or No. 299 adopted.

299. **Turned link** with adjustable end brasses. The forked end should be used where there is the greatest amount of wear.

300. **Wood connecting- or pump-rod** with wrought-iron strap ends, fitted with brasses, gibs, and cotters. Much used on mining pumps.

The shafts or rods are sometimes of cast iron of cross or T section, but are usually of a circular or flat section and swelled in the middle, similar to No. 299. See Struts and Ties, Section 101.

301. **The most usual form of shifting link** for link-reversing gear, generally got up bright all over.

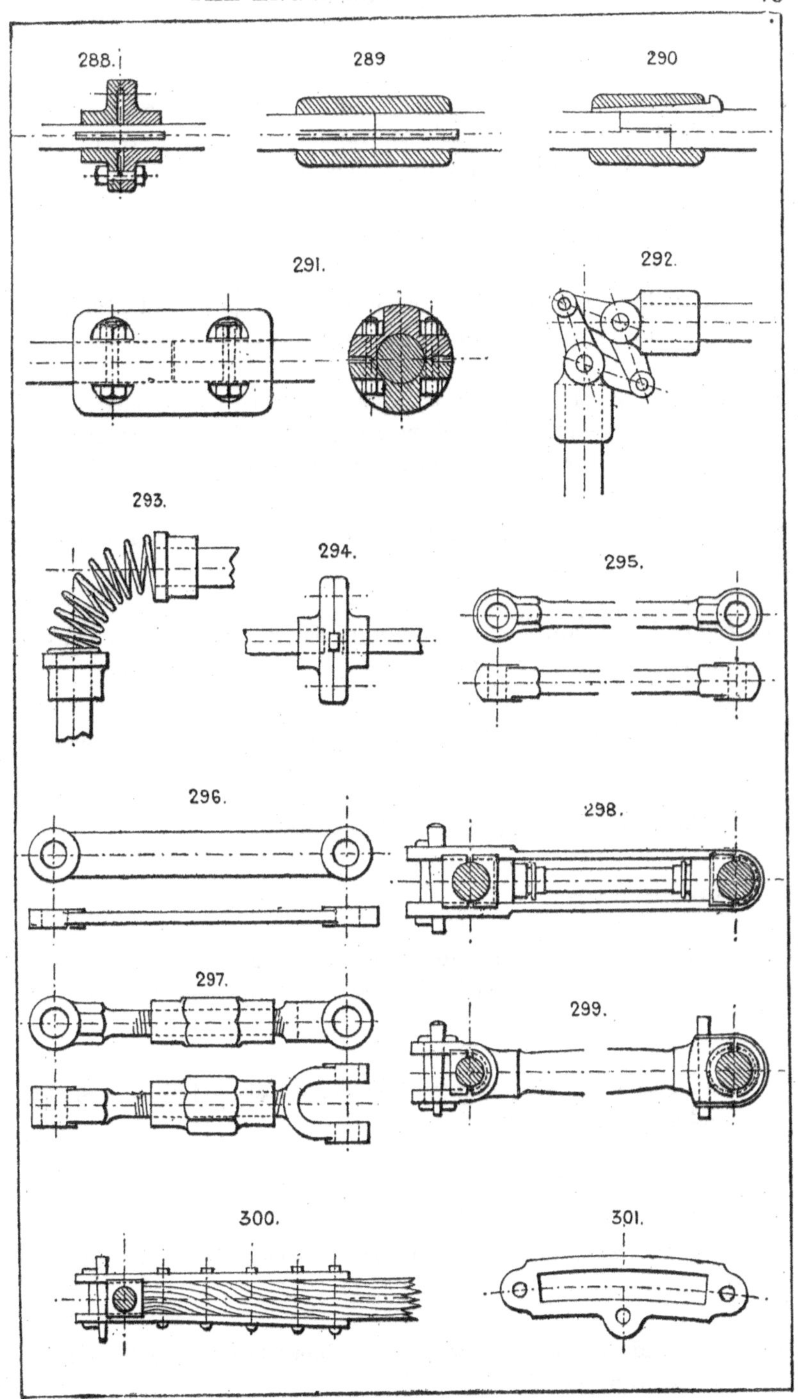
288.
289
290
291.
292.
293.
294.
295.
296.
298.
297.
299.
300.
301.

302. **Similar link,** but having the point of suspension on a side pin, fixed by screws to the link, and raised from it sufficiently to allow the sliding block and pin to pass under it.

303. **Reversed curve link.**

304. **Solid bar link,** sometimes adopted for cheapness and simplicity, the valve rod and eccentric rods having of course forked ends.

305. **Double bar link.** This is also a simple and cheap construction; the bars are plain, the rod ends single, and the block turned large enough to have a recess on each side to fit the links.

306. **Strap head connecting rod end,** with square brasses, double gibs and cotter.

307. **Strap head connecting rod end,** but with rounded end and set screw fastening for cotter.

308. **Similar to the last,** but with screw cotter adjusting device for the brasses.

309. **Solid end rod.** The brasses take out sideways.

310. **Forked end rod.**

311. **Strap end for heavy rods,** having cotter for tightening the strap to the V's in the rod end. The oil cup is often forged and turned solid on the strap, as shown.

312. **Rod end with side strap.** The brasses take out transversely by taking off the side strap.

313. **Solid end** and double set screw fastening for cotter.

314 & 315. **Solid ends for small rods.** The brasses are usually secured by a set screw.

316. **Solid end,** split with screw bolt tightening device; may be hinged as shown by dotted line.

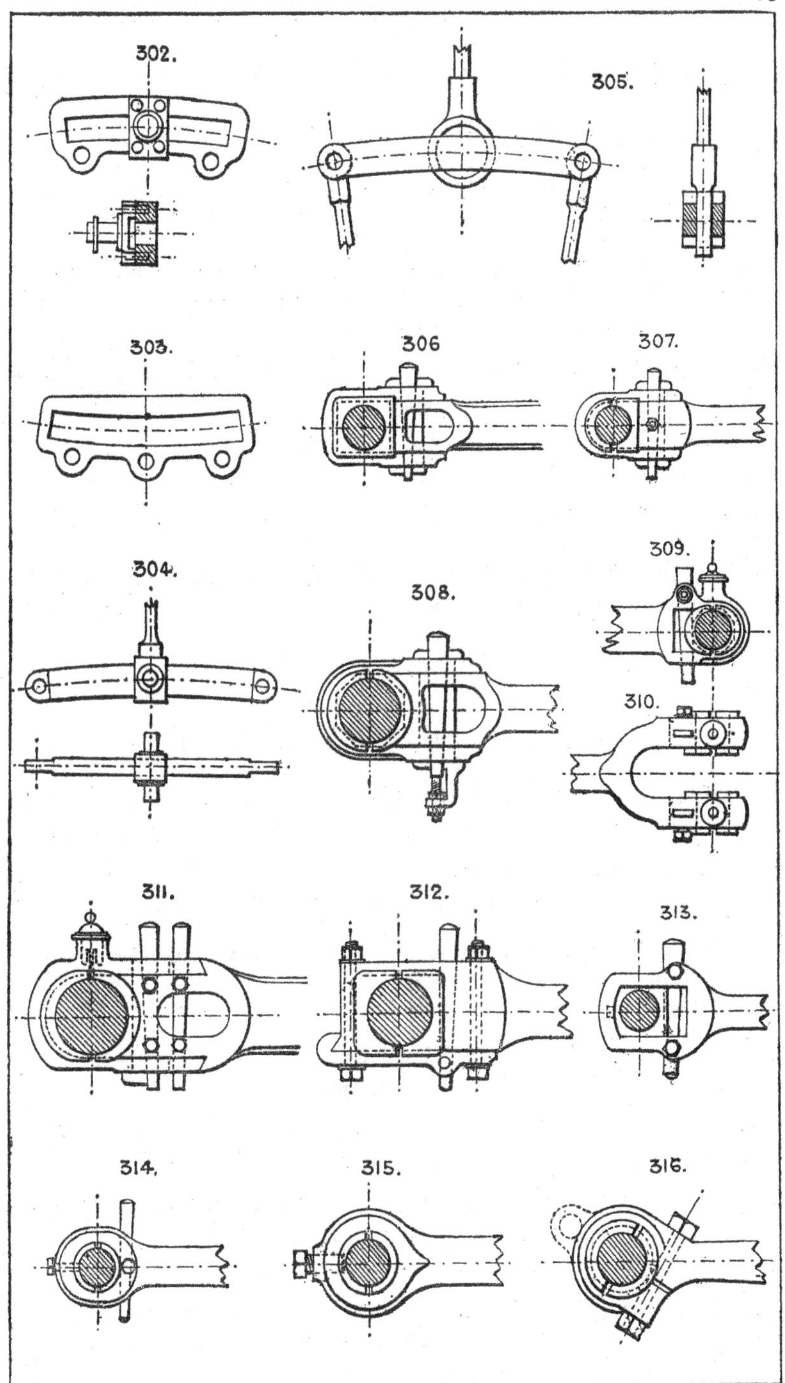
302.
305.
303.
306
307.
304.
308.
309.
310.
311.
312.
313.
314.
315.
316.

317. **Covered solid end** for crank pins, with screw adjustments for brasses.

318. **Common forked rod end,** with cap.

319. **Hook bolt attachment** for gudgeon; sometimes useful where there is thrust only.

320. **Double connecting rod,** in which the rods form also distance rods and bolts for the heads, which are in halves and fitted with brasses of the ordinary type.

321. **Marine type** of rod end, having solid end, square brasses and cap.

322. **Marine head,** in which the brasses are extended to form the central block in halves, the rod end being of T shape and bolted through the brasses and cap.

323 & 324. **Plain links.**

There are innumerable varieties of the illustrated types of heads in use, every engineer having his own design.

Section 18.—CRANES, TYPES OF.

Our object here is to indicate or suggest general design or arrangement only, from which a selection can be made to suit requirements.

325. Is **the common type of wharf crane** with fixed post, the base plate being well bolted down to a solid mass of masonry.

326. Is **also a common type of wharf crane,** but with the post revolving in a footstep and base plate; this gives a better base than No. 325.

327. **Has no post,** but a revolving frame and base plate with front and back friction rollers, and a centre pin.

328. **Post and jib in one piece,** usually of wrought iron. A balance weight is fixed at A to balance the overhanging jib.

329. **Swing derrick crane,** generally of wood. The jib turns three-fourths of a circle, and the two guys are fixed at an angle of 90° apart, and well secured by anchoring or loading.

330. **Wharf crane,** with centre tension bolt instead of crane post. In this arrangement there is a vertical tension on the centre bolt and thrust on the foot of jib.

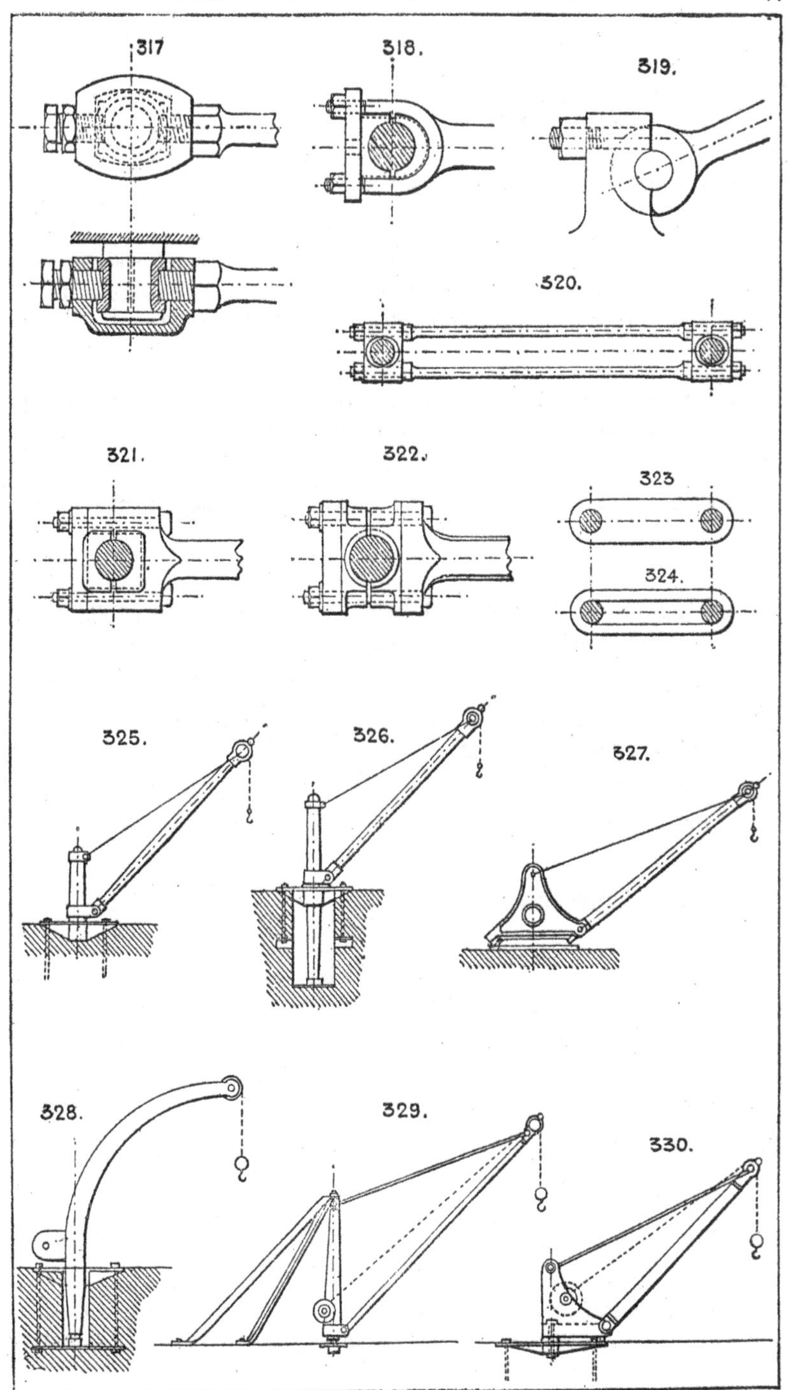
317
318.
319.
320.
321.
322.
323
324.
325.
326.
327.
328.
329.
330.

331. **Warehouse wall crane.**

332. **Warehouse wall crane, with high jib-head.**

333. **Whip crane,** chiefly used in goods sheds. The barrel is sometimes worked by an endless handrope as shown, and sometimes by a second rope and drum with a hand crank as No. 1209.

334. **Portable hand crane,** with balance weight. The balance weight can be shifted in or out to balance the load.

335. **Foundry crane,** sometimes with travelling carriage on the jib, as No. 336.

336. **Swing bracket crane** and traveller, usually formed of flat bars on edge; used only for light loads, for smiths' shops, &c.

337. **Wharf derrick,** to turn an entire circle, similar to No. 329, but employed for heavy loads.

338. **Floating derrick.**

339. **Light balance crane.**

340. **Trussed jib crane,** with centre tension bolt.

341. **Simple derrick and winch,** with two guy ropes; for temporary purposes only, and may be easily shifted about.

342. **Sheers and winch.**

343. **Tripod and winch.**

344. **Sheers with screw adjustment** to back leg. This design is adopted for very heavy lifts, such as loading heavy machinery, shipping masts, boilers, &c.

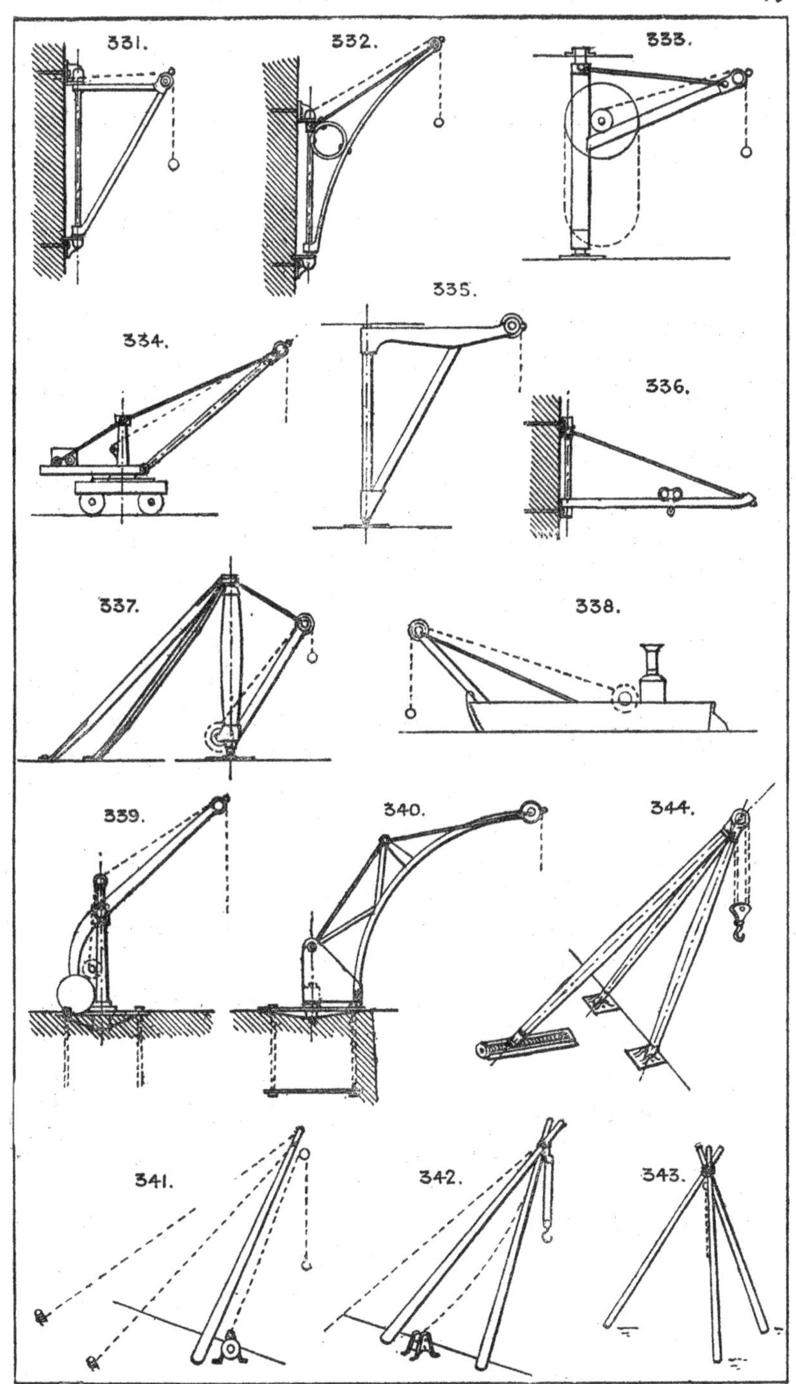
331.
332.
333.
334.
335.
336.
337.
338.
339.
340.
344.
341.
342.
343.

345. **Four-guy derrick and winch,** used for fixing columns, bases, masonry, &c.

346. **Fixed post steam crane,** for wharfs, piers, jetties, harbour works, &c.

347. **Portable steam crane,** very largely used on wharfs, piers, &c., and sometimes fitted with travelling gear in addition to hoisting and slewing motions.

348. **Wharf crane,** with fixed engine centre bolt, trussed arched jib. This is a very good type, as the ground is kept clear for goods, &c., and of course all motions, hoisting, lowering, and slewing are controlled from the crane above ground.

349. **Hydraulic wharf crane,** with fixed post. The common type universally used in docks, &c., with the ordinary form of multiplying hydraulic cylinder and chain gear; the valve for controlling its movements is operated by hand levers extending up through slots in the floor; the slewing is performed by a separate cylinder and chain gear, with a distinct controlling lever.

350. **Hydraulic short lift ram, centre crane, and traveller,** employed chiefly to raise the ingots out of the casting pits of Bessemer steel works. The ram is of course subject to severe cross strains, and many designs provide an overhead guide or support for the ramhead.

351. **Automatic balance crane,** portable or fixed; the position of the fulcrum varies with the load.

352. **Steam multiplying cylinder crane,** in which the ram is forced out by steam pressure, acting either directly or by an intervening body of water.

353. **Breakwater swing crane.**

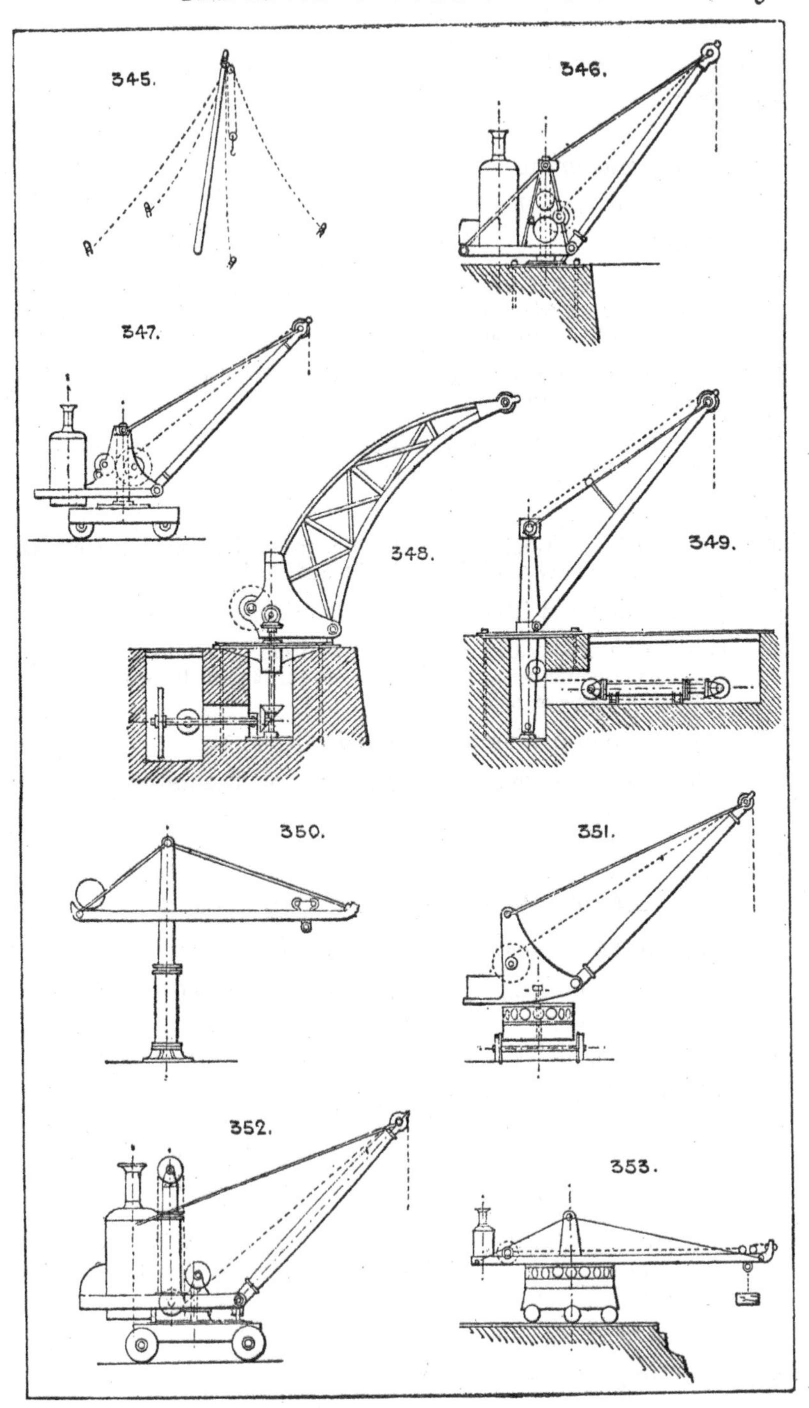
345.
346.
347.
348.
349.
350.
351.
352.
353.

354. **Overhanging travelling crane,** for use on breakwaters, &c.

355. **Overhead hydraulic travelling goliath,** to span a railway; has slewing motion and a balanced jib.

356. **Single rail crane** with top guide rail.

357. **Overhead traveller** on gantry.

358. **Goliath.**

359. **Steam overhead crane,** with carriage to span a railway. Largely used on dock wharves, &c., as they give a high lift and do not encumber or encroach on valuable quay space.

360. **Hydraulic cylinder post crane;** sometimes adopted instead of the type No. 349.

361. **Heavy hydraulic crane,** with suspended cylinder; employed for work of the very heaviest class.

362. **Ship's davit.**

363. **Balanced jib post crane,** no tie rod. The weight must be sufficiently heavy to balance the jib and load.

364. **Hydraulic strut jib crane.** The load is raised by raising the jib.

365. **Overside dock crane,** for discharging from ships into barges. The overhang being very great in this design, it must be provided with a heavy frame or balance weight.

366. **Wagon tip crane,** for loading vessels.

367. **Double sheave** 4 to 1 purchase for crane jib. See also Section 69.

Section 19.—CONVEYING MESSAGES.

Messages can be conveyed by—

1. **Speaking tube;** for distances up to say 300 feet, $\frac{3}{4}''$ to $1''$ bore tubes.
2. **Telephone;** any distance.
3. **Telegraph;** any distance.
4. **By signals**—(a) Wire or cord and bell; (b) sight signals, such as the semaphore, lamps, heliograph, flags, and other devices; (c) by sound, such as a bell, trumpet, siren, whistle, &c. See Section 108.
5. **By pneumatic despatch:** that is, by forcing a piston carriage containing the message or small parcel through a tube by compressed air.
6. **Carrier pigeons.**

368. **Signalling dials** and **bevel gearing.**

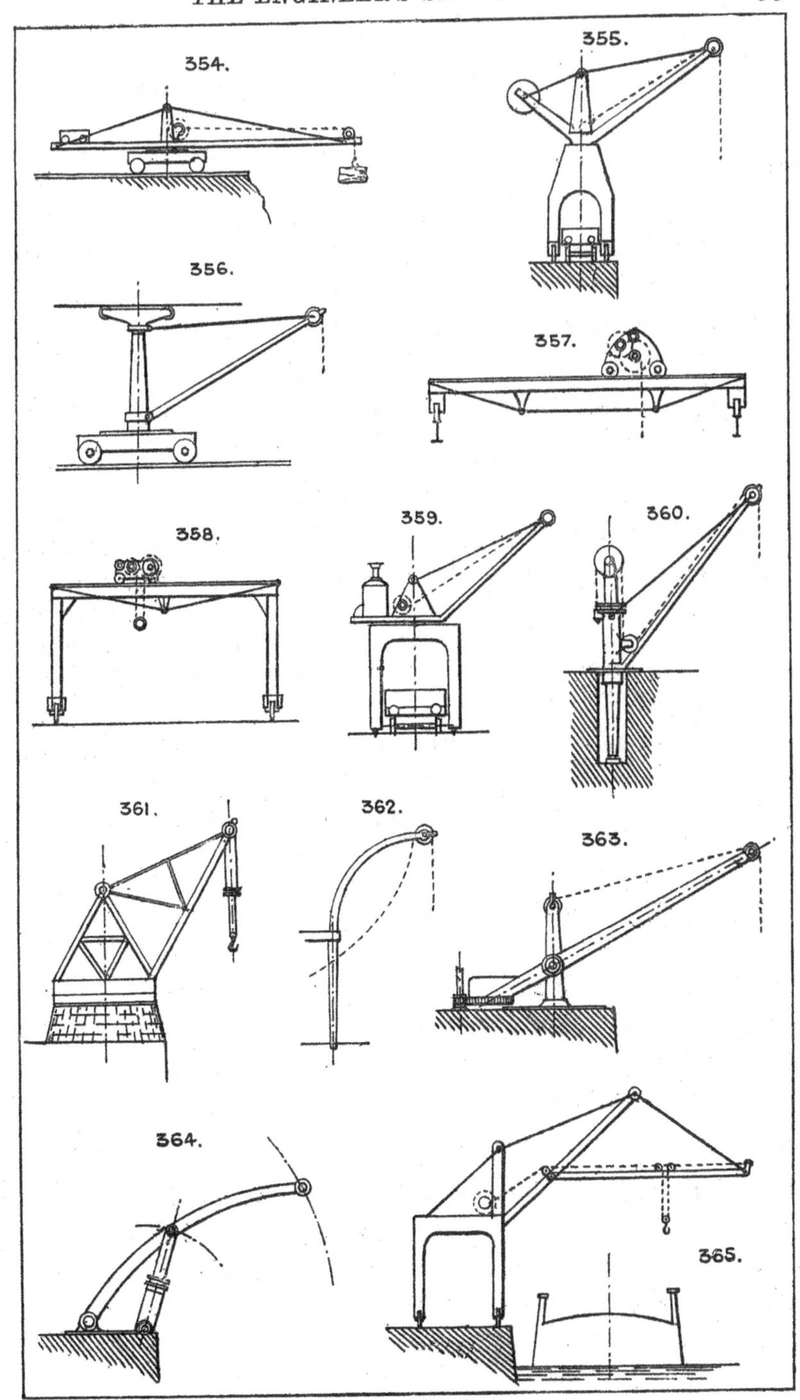
354.
355.
356.
357.
358.
359.
360.
361.
362.
363.
364.
365.

Section 20.—COMPENSATING AND BALANCE WEIGHTS.

It is of the utmost importance that every revolving or reciprocating part of any machine should be as nearly as possible balanced, to obtain smooth running with the least amount of wear.

The following are types of the most important devices and their applications:—

369. **Balanced lever,** having a sliding cheese or ball weight fixed with a set screw.

370. **Balanced cage of hoist.** It is usual to *over-balance* the cage to divide the work between the up and down journeys in hand-power lifts to assist the load; but in power and hydraulic lifts the cage is under-balanced so as to descend when empty.

371. **Hydraulic balance lift,** in which the dead or constant load of cage and ram are balanced by a loaded piston in a supplementary cylinder; to raise the loaded cage the pressure water is admitted to the upper side of this piston. Many varieties of this type are in use; see Ellington's, Johnson's, Stevens and Major's, Waygood's, and other patent lifts.

372. **Variable volute compensating balance** for revolving shutters, blinds, curtains, &c., to maintain an even balance in all positions.

373. **Variable compensating balance** for hydraulic lift rams, to compensate for loss of immersion of the ram as it ascends (Berly's patent). See also Stevens & Major's patent, where bell-crank levers and weights are employed instead of loaded chains. See No. 383.

374. **Balanced fly-wheel.** For balanced cranks, see Nos. 172 & 173.

375. **Increasing balance by sections,** lifted at intervals as the chain rises.

376. **Balanced riveting machine.** See Tweddell's patents.

377. **Variable lever balance.** For balanced cranes, see Section 18.

378. **In deep lifts,** to balance the weight of chain or rope, it is **made endless.**

379. **Another method.** The loose chain hung from cage is of the same weight per foot as the lifting chain.

380. **Balance weight on a screw arm** for adjustment, employed on weighing machines.

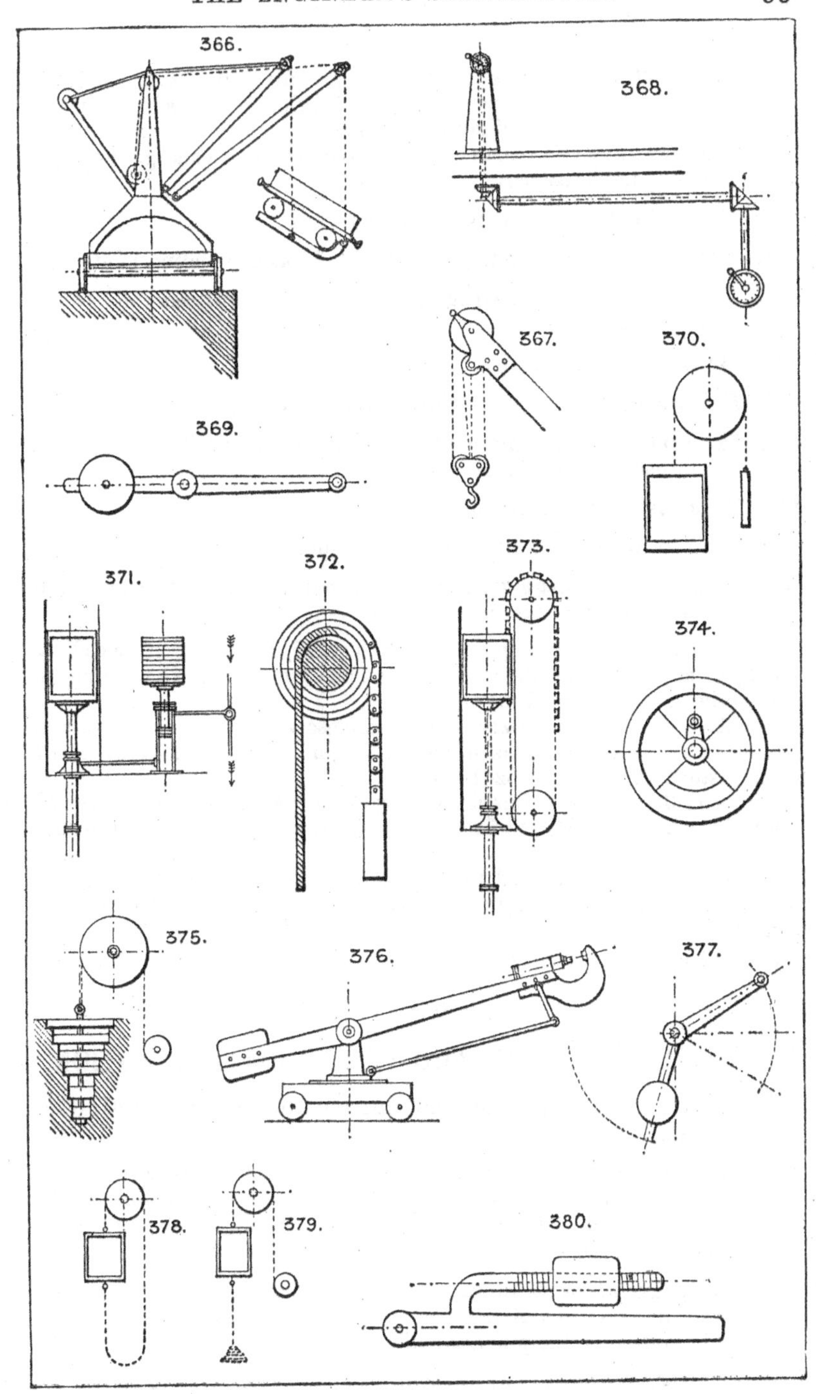
366.
368.
367.
370.
369.
371.
372.
373.
374.
375.
376.
377.
378.
379.
380.

381 & 382. **Worthington's compensating air cylinders**, employed on direct-acting horizontal pumps, working expansively, in lieu of fly-wheel. The oscillating or vertical cylinders are air or spring pistons, absorbing power the first part of stroke and giving it out during the latter part.

383. **Variable balance weight** by bent lever.

384. **Variable balance weight** by double links and sliding joints.

385. **Dawson's compensating governor.** See *Mechanical World*, August 25th, 1888.

386. **Balanced hinged doors.**

387. **Balanced sashes,** or vertical sliding doors.

388. **Method of balancing a bloom** in charging or withdrawing from a furnace, or any similar use.

389. **Balance** for link motion.

390. **Weight** to keep a cord or rope in tension.

391. **Mode of balancing two sliding doors** so that they rise and fall at proportionate speeds.

Hoisting and winding engines (see Nos. 1222, 1223) are balanced by having an ascending and descending cage, and two ropes, one winding on as the other winds off the drums.

Double cage hoists similarly balance themselves. Heavy slide valves, and other reciprocating parts of steam engines, are balanced by small steam pistons. See Nos. 1652–1655.

Foot treadles, when required to always stop at a point off the dead centre, have a balance weight fixed to fly-wheel, at right angles to the dead centre.

A water tank is often used to serve as a counterpoise, or balance, and may be made variable by varying the quantity of water by a siphon or other device.

For Balanced Valves, see Section 89.

Section 21.—CIRCULAR AND RECIPROCATING MOTION.

392. **The ordinary type of piston-rod and crank motion** as universally used.

393. **Watt's substitute for the above,** or "sun-and-planet" gear. Note that the crank shaft revolves twice for each double stroke or revolution of the engine. The crank being a loose link only, the planet wheel does not revolve.

394. **Epicycloidal parallel motion and crank.** The pinion is one-half the diameter of the wheel on pitch line, and the connecting pin is fixed on the pitch line of pinion.

395. **Bernay's patent crank motion**; radius of crank = stroke × ·25.

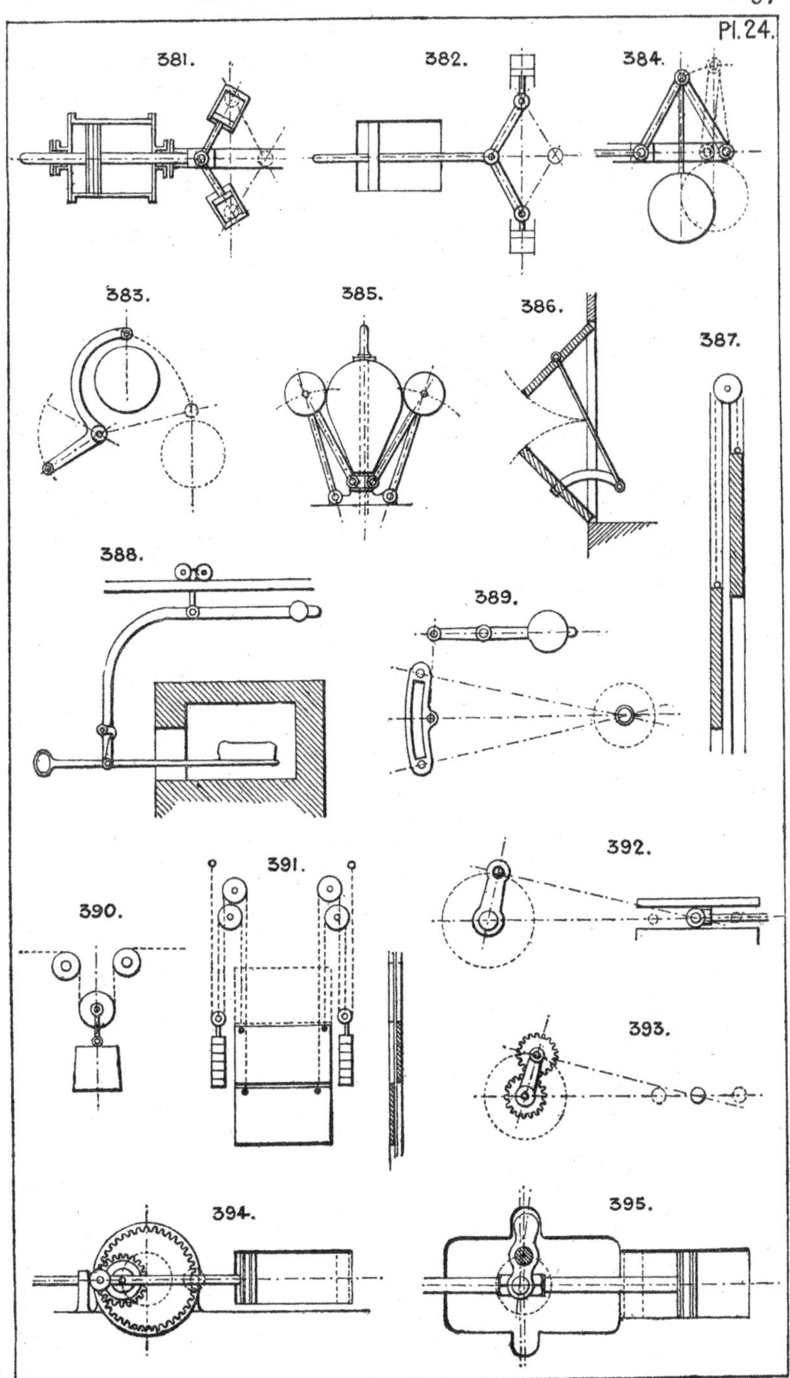
Pl. 24.
381.
382.
384.
383.
385.
386.
387.
388.
389.
390.
391.
392.
393.
394.
395.

396. **Slot and crank motion.** The pin usually runs in a sliding block.

397. **Segment pinion and double rack motions.**

398. **Rack and pinion.** The pinion is sometimes made so as to be driven on one stroke and run loose on the other, by a clutch or ratchet motion, such as Nos. 1135, 1178, or their equivalents.

399. **Hydraulic multiplying gear.** See also Section 42.

400. **Slotted crosshead and disc crank.** The pin runs in a sliding block in a groove in the covered crosshead.

401. **Stannah's patent,** works vertically; the fly-wheel centre A oscillates on the end of a link B, allowing the crank pin to run in a straight line.

402. **Screw and fly nut.** May be made to produce continuous rotary motion by fitting the nut with a clutch motion similar to 1135 or 1178, so as to grip the wheel only on one stroke.

403. **Friction gear**; the pinion is driven by the reciprocating rod and runs loose on the out stroke, the weighted lever with roller giving frictional grip on the in stroke.

404. **Lever and roller crank pin.**

405. **Treadle motion,** with cord and spring. For continuous rotary motion the pinion must be fitted as described with No. 402.

406. **Ball and socket crank motion.** The crank pin is always horizontal.

407. **Segment lever,** with cord and pulley.

408. **Double geared cranks,** used for driving rotary blowers, &c.

409. **J. Warwick's patent**; circular motion converted into reciprocating by a diagonal sheave grooved as shown; the crank arm centre is in line with the centre of the sheave, as shown in dotted lines.

410. **Rolling sectors,** with thrust motion to crank pin. Used in Outridge's box engine with double pistons; this gives a constant rectilinear thrust to the crank pin at all points in the stroke, and no part is in tension.

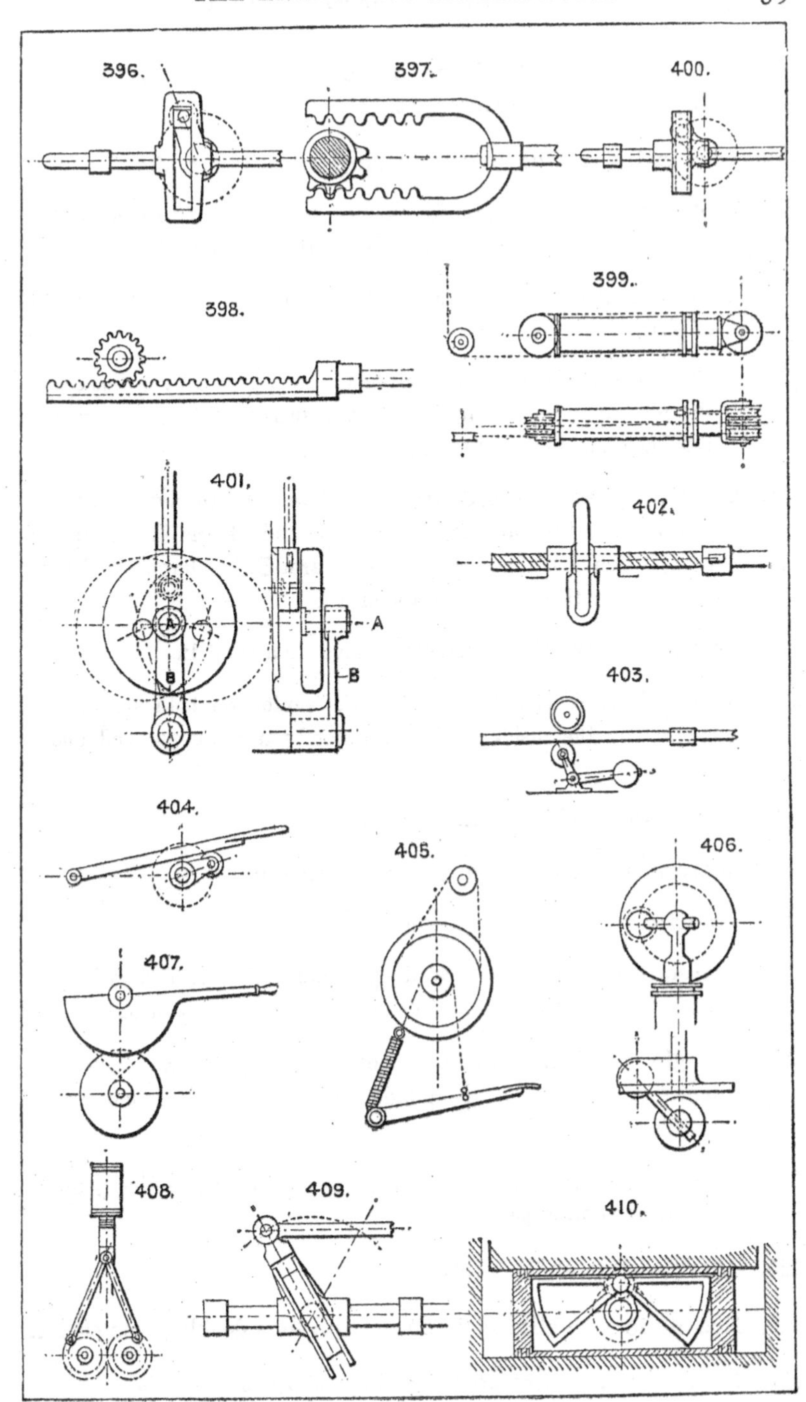
396.
397.
400.
398.
399.
401.
A
B
A
B
402.
403.
404.
405.
406.
407.
408.
409.
410.

411. **Weight and multiplying pulleys,** used for clock motions, driving any light machines, &c.

412. **Oscillating clutch arm and ring,** silent feed motion.

413. **Slot and roller motion** for crank. The crank pin has a friction roller, which runs in a covered slot in the crosshead.

414. **Trammel gear;** one revolution of the wheel to two double strokes of piston.

415. **Segmental vanes** (in a semicircular case), driven by a disc crank and pin, giving motion by links to two arms fixed to the two vanes. Used as a pumping or blowing machine.

416. **Circular into reciprocating motion** by revolving arm A carrying the two pinions, the point at end of arm B describes a vertical line four times the length of arm B, the large wheel C is fixed, and motion is given to the arm B.

417. **Trammel gear;** the slotted cross moves in a right line.

418. **Slot link and treadle,** driving the pinion on both strokes by friction on the inside of link alternately at the upper and under sides.

419. **Chain and roller treadle motion.**

420. **Reciprocating wheel and crank motion.**

421. **Velocipede pattern** foot treadle.

422. **Double crossheads,** separated by distance rods so arranged as to allow the crank and connecting rod to work between them.

423. **Mangle rack and pinion reciprocating gear.** The rack moves in a right line, the pinion working round it by moving up and down the slot.

424. **Mode of connecting an oscillating lever** by a sliding joint to any reciprocating part.

425. **Suspended treadle motion.**

426. **Eccentric and sliding bush motion** for a double piston engine.

427. **Rocking lever motion** by gearing and a tied crank pin.

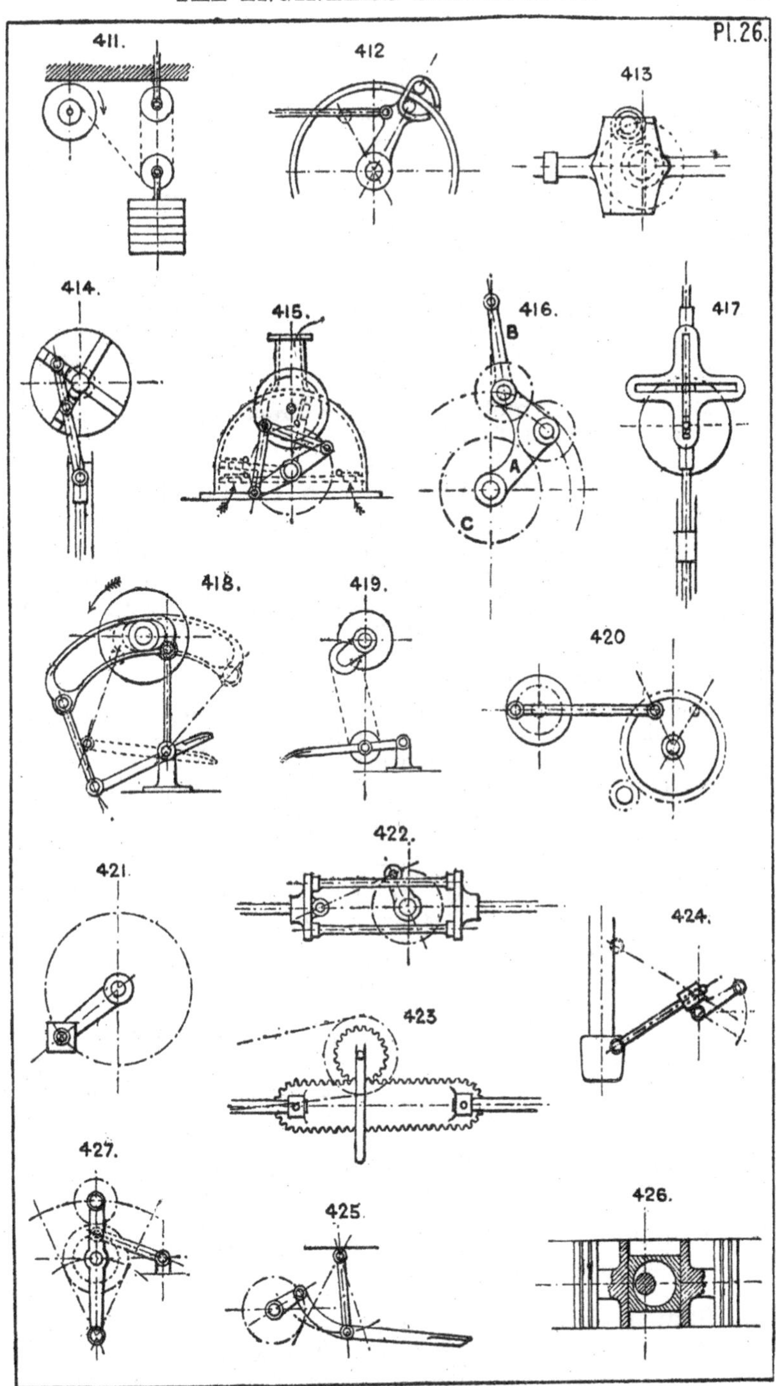
Pl. 26.
411.
412
413
414.
415.
416.
B
A
C
417
418.
419.
420
421.
422.
423
424.
425.
426.
427.

428. **Crank pin and slotted lever** for giving a variable speed to the connecting rod.
429. **Side gudgeon crank motion.**
430. **Bell crank and disc crank motion,** the bell crank centre having horizontal as well as vertical movement.
431. **Worm wheel and screw reciprocating motion** by means of a tied crank pin.
432. **Treadle, cord and pulley crank motion.**
433. **Circular into reciprocating motion,** or *vice versâ.*
434. **Another form of sun-and-planet gear.** The ring is stationary, and the bush on which the planet wheel revolves is slotted to fit the ring; the planet wheel is fixed to the connecting rod end.
435. **Bent shaft and arm motion.**
436. **Reciprocating motion** by a return thread screw and lever. See also Sections 62, 31, and 74.

Section 22.—CONCENTRATED POWER.

Multiplication of power by great reduction of speed is accomplished by the following devices, and various obvious modifications of them. Ordinary methods comprise—Gearing (see Section 84), the screw or compound screws (see Section 78), and the wedge and lever (see Section 53). By differential screws, Nos. 1379, 1380.

437. **Compound lever.**
438. **Double toothed-cam and lever combination.**
439. **Double lever and link motion,** with increasing pressure. The strains are self-contained, and this plan is very suitable where an increasing pressure is required.
440. **Lever and toggle motion** (see Section 63). Many variations are in use for stone breakers, &c. See Section 13.
Knapping toggle motion. See Nos. 269, 251.

Section 23.—CONVEYING MOTION TO MOVABLE PARTS OF MACHINERY.

Motion may be conveyed to such parts of a machine as require to be movable, or to distinct machinery which has no fixed location, by the following means :—

441. **Is an endless rope or other round section belt,** kept tight in any position in the plane of the driving pulley by a weighted pulley. In this plan the machine can be moved to any position in the plane of the driving pulley, the weighted pulley taking up the slack of the belt.
442. **Flexible shaft for light driving.** It admits of considerable flexure, and is useful for drilling and similar incidental driving purposes in difficult positions.
443. **Radiating arm and belt.** The movable machine can be driven at any point in circumference of the circle described by the arm head.
444. **Similar plan,** but driven by bevil gear instead of belt.
445. **Bevil gear and feather shaft.** The movable machine having a travel in a straight line the length of the shaft as well as a radiating motion

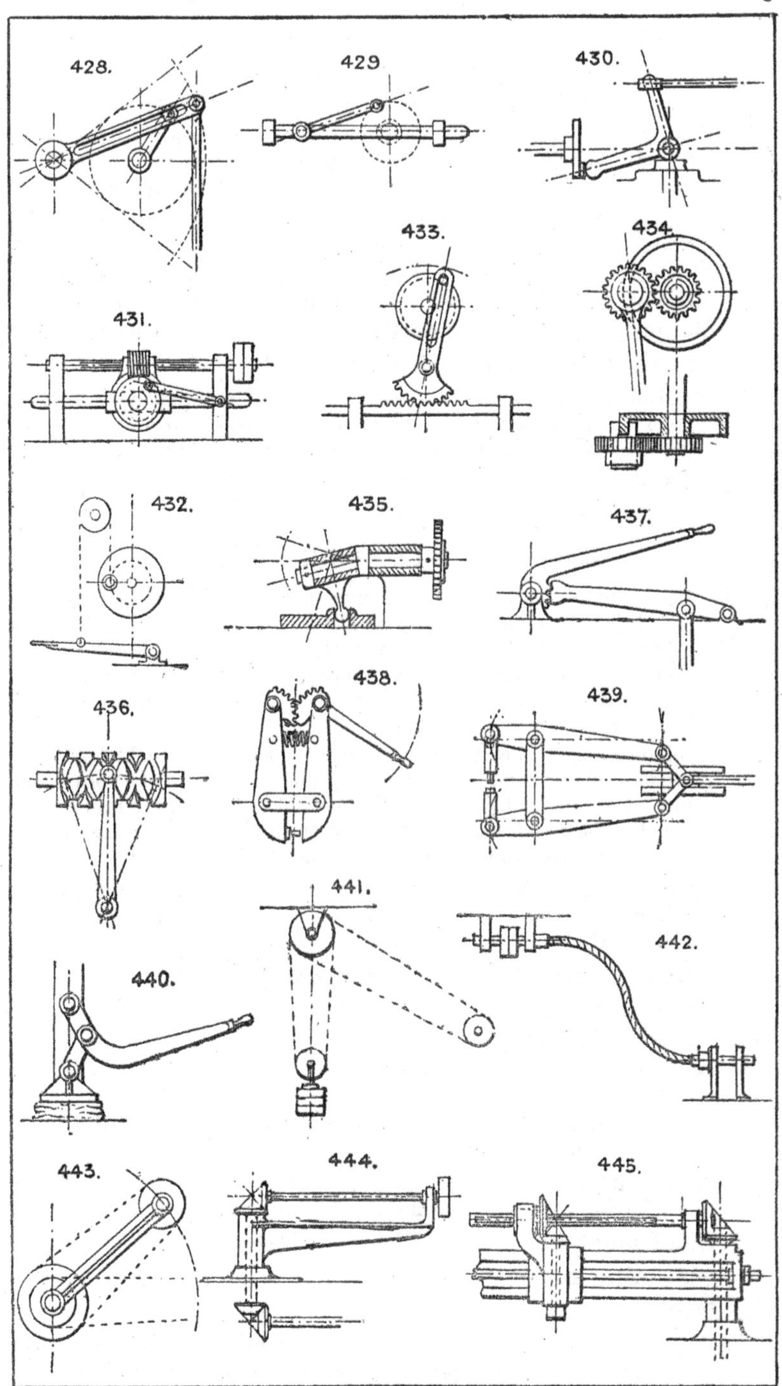
428.
429
430.
431.
433.
434.
432.
435.
437.
436.
438.
439.
441.
442.
440.
443.
444.
445.

446. **Screw and worm wheel gear,** for the same purpose as 445.

447. **The driven wheel** A has a limited travel up and down the slot, the idle wheel B being kept in gear by the link suspension.

448. **Idle wheel and slot.** A common device for changing direction or speed in driving gear by connecting or disconnecting it with intermediate gearing between a fixed driving and a driven shaft.

449. **Parallel motion radiating driving device,** with a limited vertical travel and a radial motion.

450. **Motion by belt** is conveyed to a driven shaft having a radial motion in a vertical plane. Used for light drilling, emery wheels, &c.

451. **Steam or hydraulic radiating arm** and cylinder device.

452. **Central cylinder** and radiating lever motion.

453. **Jointed radiating arms,** with belt gear for conveying motion from a central spindle to one having a travel covering any point within a circle of the extreme radius of the jointed arms.

See also Nos. 348, 349.

Section 24.—CUTTING TOOLS.

Besides the ordinary cutting tools in use in the workshop, such as the chisel, gouge, plane, saw, drawknife, scissors, shears, scythe, and others, and which do not properly belong to machine devices, there are others, some of them mere modifications of the ordinary tools that are sometimes needed in the design of machines, and are illustrated here.

Other appliances are—Shears: see the ordinary shearing machines, bookbinder's shears, and other modifications. In some the shears are hinged at one end, in others the movable blade moves either with equal or unequal motion at either end by cam or crank motion (see 462). Lawn mower revolving cutters (see 486).

454. **Pipe cutter,** with V-edged cutting roller. Sometimes 3 cutting rollers are used.

455. **Cutting discs,** used for paper, sheet metal, &c.

456. **Slitting discs,** for cutting sheets into strips.

457. **Revolving cutter head,** for moulding, tenoning, and numerous wood working uses.

458. **Hollow revolving cutter head,** for rounding wood rods, broom handles, &c. See also No. 488.

459. **Reaping machine cutters.** A series of scissor-shaped knives, one set fixed and the other reciprocating.

460. **Wire cutter discs,** one fixed, the other attached to the hand lever, and having corresponding holes of various sizes in both discs.

461. **Chaff machine,** with revolving shear blades.

462. **Guillotine shears.**

463. **Milling cutters.**

464. **Tubular machine cutter** for wood working; easily sharpened, and can be revolved to present fresh cutting edges to the work.

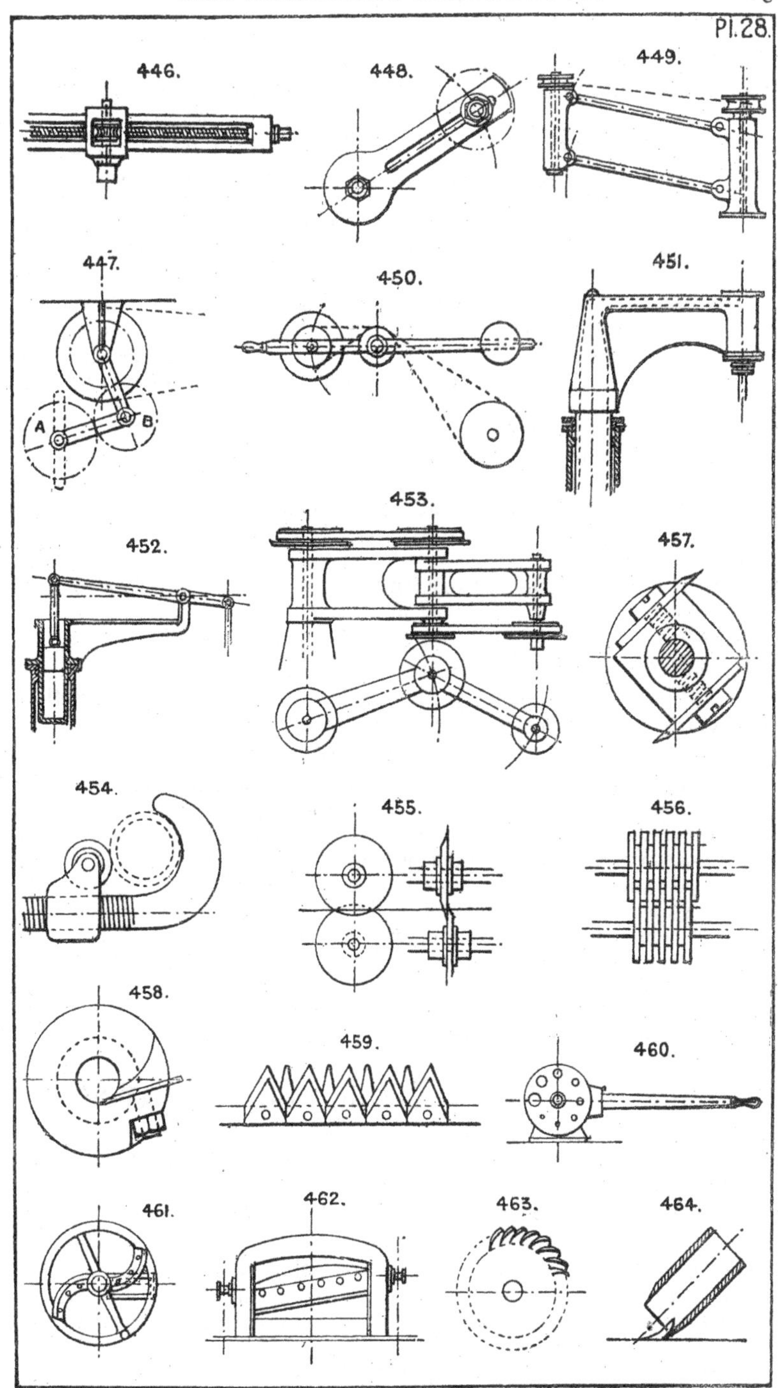
Pl. 28.
446.
448.
449.
447.
A
B
450.
451.
452.
453.
457.
454.
455.
456.
458.
459.
460.
461.
462.
463.
464.

465. **Fret saw** or jigger.
466. **Three-cutter tube shears,** with worm gear motion.

Section 25.—CONDENSING AND COOLING APPLIANCES.

Their uses generally are to condense steam, to cool heated gases, air, or articles of food requiring a low temperature; distilling, and other purposes. For cooling purposes, compressed air machines are in most demand. The air is compressed in a cylinder, then cooled to ordinary temperature again in a surface condenser, such as No. 468, and then expanded into the cooling chamber, through a cylinder and piston, the expansion reducing its temperature usually to 10° or 20° below zero. Other cooling appliances are ammonia machines, fans, and blowers of all kinds, punkahs, or waving fans, freezing mixtures, &c.

467. **Gravity condenser.** The pipe should be 34 feet high or more, in which case no air pump is required, as the condensed steam and air are discharged below. In place of the pipe an air pump and foot valve are required, and are commonly used, as it is seldom convenient to have a vertical pipe 34 feet long with a water supply at the top.
468. **Surface condenser, multitubular.** The steam may be led into the tubes, and the water around them, or *vice versâ*.
469. **Worm, or coil condenser,** chiefly used for distilling.
470. **Still condenser for essences,** spirits, &c.
471. **Condensing chambers for gases, &c.** Horizontal or vertical.
472. **Wimshurst's condenser,** requires no air pump.
473. **Another form of ejector condenser.**
474. **Tray cooler, or condenser;** a series of water trays supplied from a tank above.

See Morton's ejector condenser, which requires no air pump; Hayward's exhaust condenser, which employs the water in suction pipe of a pumping engine to condense the steam. See Messrs. Tangye's list. Water tube cooling coils are used for tuyeres and other hot surfaces.

Air-compressing and gas engine cylinders are water jacketed to carry off the heat of the compressed air or gas. Cooling by exposing a large surface to air is sometimes employed for exhaust steam on tram car engines &c., the apparatus consisting generally of numerous wrought-iron tubes or coils.

Section 26.—CONCENTRATING AND SEPARATING.

Sifting, riddling, and screening are treated of under Section 72. For concentrating ores many methods are in use, of which the water processes are the most important.

475. **Circular revolving concentrating table.** The lightest particles are discharged over the edge, and the heaviest remain in the centre.

The ordinary magneting machine, for separating particles of iron or steel from mixed borings, &c., consists of a series of magnets drawn through the material, and then through fixed brushes, which brush off the iron particles adhering to the magnets.

476. **Separating dust from grain, &c.,** by a current of air driven through the stream of material as it falls from hopper to hopper. See also Nos. 1268, 1270.

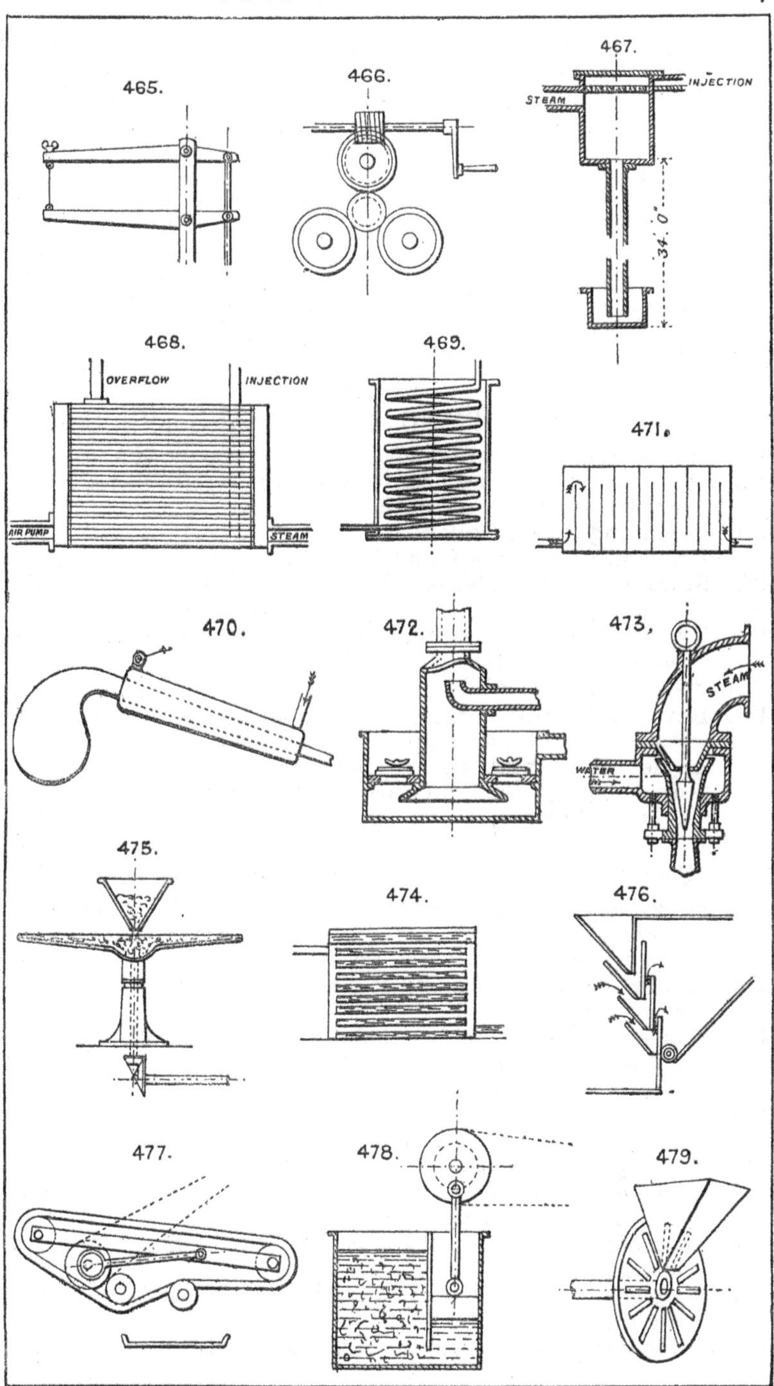
465.
466.
467.
INJECTION
STEAM
34' 0"
468.
OVERFLOW
INJECTION
AIR PUMP
STEAM
469.
471.
470.
472.
473.
STEAM
WATER
475.
474.
476.
477.
478.
479.

477. **Ore concentrator**; consists of an endless rubber belt with flanges (see No. 1082), having a slow longitudinal motion, and a rapid shaking motion, either sideways, as in the "Frue Vanner," or endwise, as in the "Embrey" concentrator; a stream of water runs over the ore, the heavy particles settle on the belt, and the mud is washed off.

478. **Jig** for separating ores by motion of a piston in water, the heavy parts settle to the bottom and the light parts are removed at the top.

Filtration through various substances—as sand, charcoal, calcined ores, &c., is employed to separate suspended matter from liquids.

Separation by subsidence in a tank, similar to No. 1571, is employed for lime, &c.

Chemical deposition and evaporation are necessary in many cases.

Section 27.—CHOPPING, SLICING, AND MINCING.

479. **Has a disc cutter** with radial knives and slots; used for roots, &c.
480. **Disc cutter**, with small knives wedged in separate holes, through which the cuttings escape in shreds.
481. **Revolving cutter rollers.**
482. **Hand mincing compound knife.**
483. **Spiral tapered revolving cutter**, in a conical case, having projecting knives on its interior. The type of the common mincing machine.
484. **Two or more rectangular cutters**, with vertical reciprocating motion in a revolving pan for mincing.
485. **Single roller revolving cutter** machine.
486. **Revolving spiral cutters**, as used in the common lawn mower, in connection with a fixed straight knife or shear blade.
487. **Apple slicer and corer** (cutter for). The apple is passed down through the cutter and divided into sectors and central cylindrical core.

Section 28.—CHUCKS, GRIPS, AND HOLDERS.

Common devices for gripping articles comprise the ordinary vice, tongs, pincers, pliers, joiners' handscrew, cramp, bench screw, parallel vice, instantaneous grip vice, &c.

488. **Hollow chuck**, with radial knives, for rounding wood rods. See also No. 458.
489. **Barber's patent** grip for shanks of drills, brace bits, &c., having square taper shanks.
490. **Collar grip** and bolt, or set screw.
491. **Cone and screw lever grip**, with two or more jaws; with two jaws only it serves as a small vice.
492. **Taper grip** for vices.
493. **Tool box**, for lathes, planing machines, &c., with central revolving tool post and set screw.
494. **Tool box**, with two tool stocks and set screws sliding in T grooves in the slide rest.
495. **Tool box**, with clamping screw and plate, which can be revolved to any angle.
496. **A modification of 495**, the tool being secured by set screws in the clamping plate

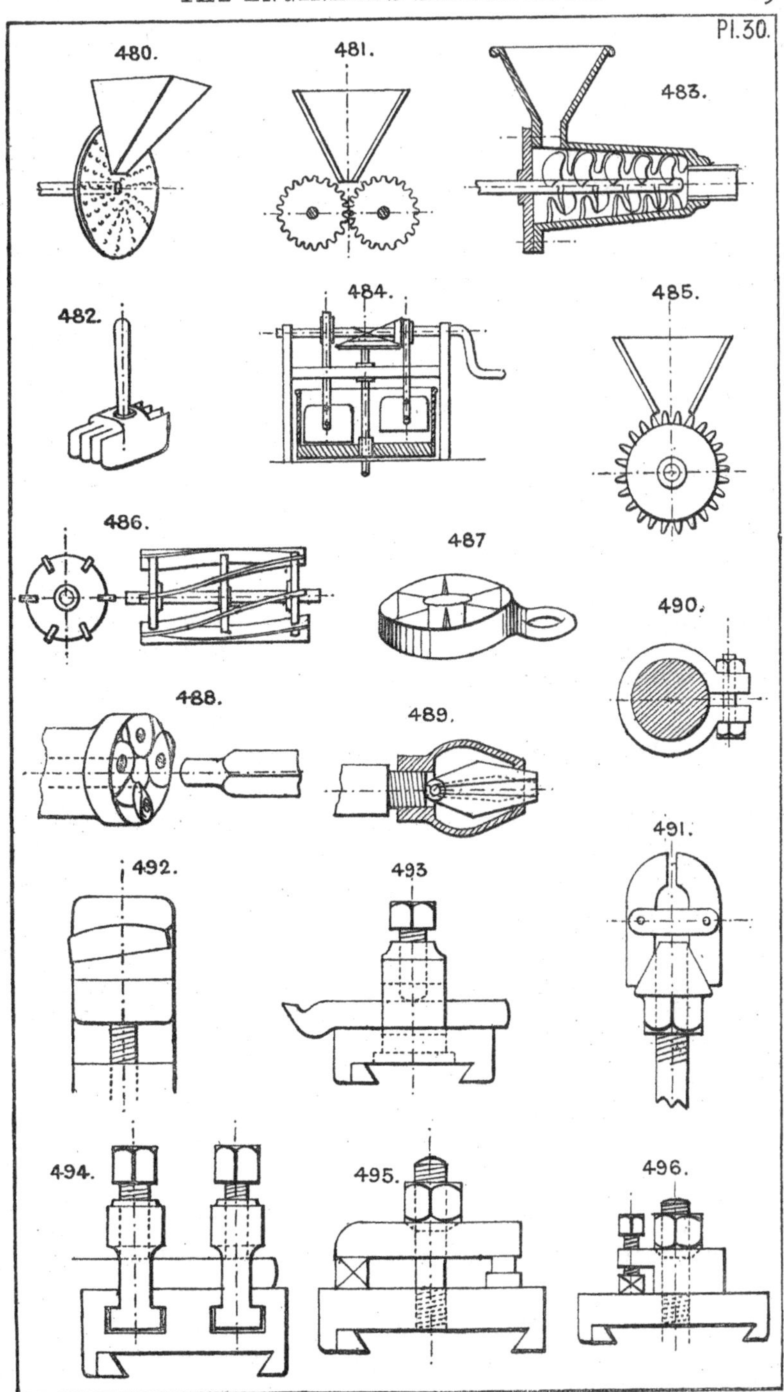
Pl. 30.
480.
481.
483.
482.
484.
485.
486.
487
490.
488.
489.
491.
492.
493
494.
495.
496.

497. **Rail grip** for holding a crane, car, &c., down to its railway.

498. **Cam-lever rail grip** for safety gear on inclines; this is usually thrown into action by a spring released by the breakage of the hauling rope.

499. **Cone centering grips** for machine tools.

500. **Hinged clamp,** with screw and nut.

501. **Fitter's clamp or cramp.**

502. **V grip vice** for round rods and tubes.

503. **Lathe carrier,** for round rods, spindles, &c.

504. **Bench cramp;** employed to hold down to the bench work operated upon; the bench has a series of holes bored in it to receive the vertical leg of the cramp.

505. **Grip tongs,** used for draw benches, &c.

506. **Split cone expanding chuck** for rods, &c.; the centre cone is split into three or four parts, and the screwed ring or collet contracts the split cone upon any cylindrical article inserted in the central aperture.

507. **Le Count's patent expanding mandril,** with cone and three sliding feathers which are fitted into dovetail grooves in the conical mandril.

508. **Bell chuck** and set screws for lathes.

509. **Three jaw grip,** or stay bearing.

510. **Pipe tongs,** self gripping; there are several modifications in use.

511. **Paper grip,** used for holding sheets of paper; released by striking a stop at any point in the travel of the machine.

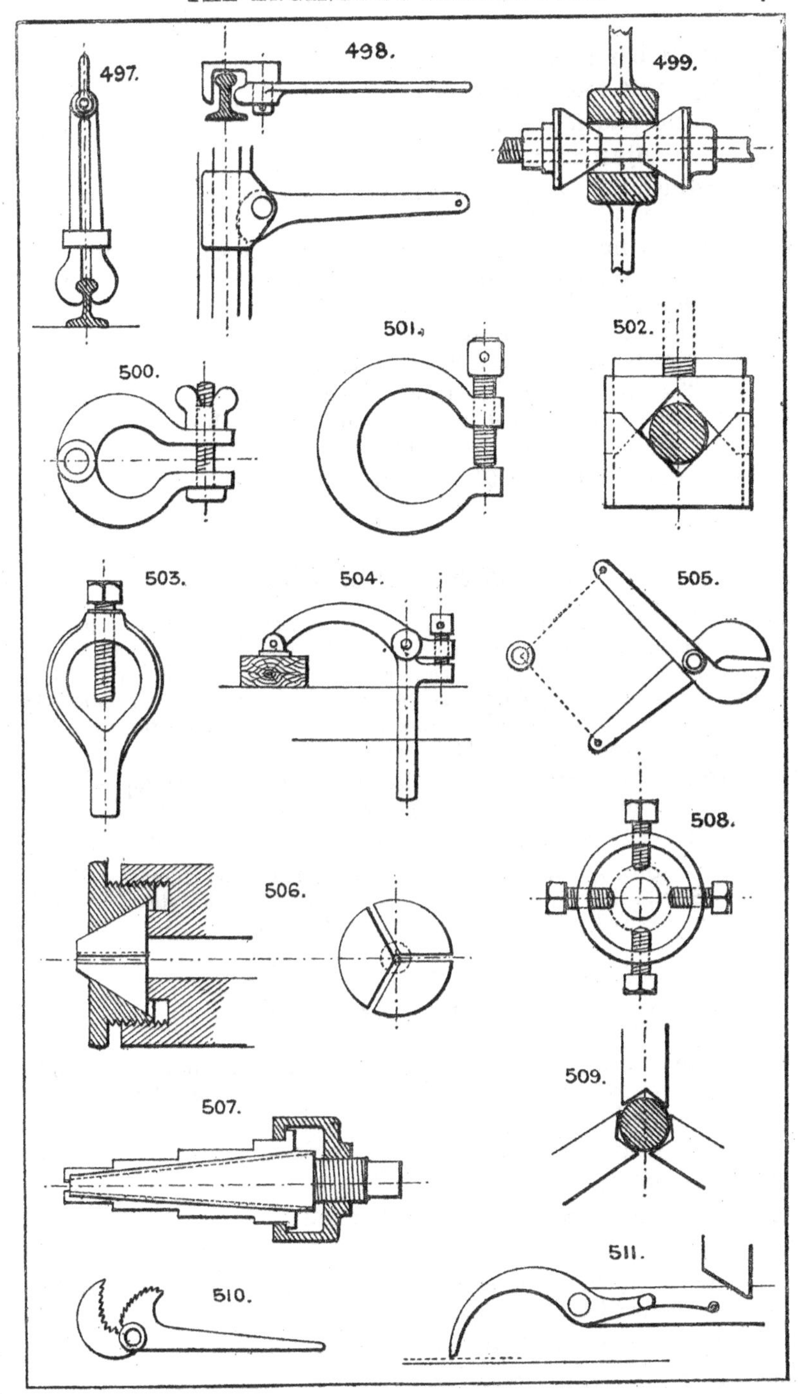
497.
498.
499.
500.
501.
502.
503.
504.
505.
506.
507.
508.
509.
510.
511.

512. **Split bar grip,** or tool holder.

513. **Eye-bolt tool holder.**

514. **Hand pad** for holding small tools.

515. **Self-adjusting jaws** for round articles.

516. **Adjustable gripping tongs** for lifting heavy stones, boxes, &c. See also No. 761.

517. **Revolving tool post, or head,** to carry a variety of tools, each being required in use in a certain order, as in special repetition turning work.

518. **Double screw gripping tongs.**
See also No. 944, 912, 918, 917, 919, 923.

The ordinary three or four-jaw chucks, wood chucks with centre screw or fork, and numerous varieties of self-centering chucks, are well known. See tool makers' lists.

Spindle grips, Nos. 917, 918, 979.

Section 29.—CUSHIONING.

For checking the impact of a blow, or more generally the momentum of a heavy moving part of a machine. The devices in use comprise (*a*) springs, see Section 80; (*b*) air cylinder, see No. 1480; (*c*) pistons driven by *elastic* fluids, such as steam and air, can be cushioned by imprisoning a portion of the fluid at each end of the cylinder; (*d*) brakes of various kinds, see Section 5.

519. **Hydraulic cushion.** The descending ram, by its tapered end, closes gradually the discharge outlet for the water.

Hydraulic buffer stops are constructed on this principle.

520. **Cushioning device,** at the upper end of a steam-hammer cylinder. Should the piston pass the exhaust holes, the steam above is imprisoned, and checks the piston without shock.

Section 30.—DRILLING, BORING, &c.

Besides the ordinary tools in use, as gimlets, bradawls, pin and brace bits, augers, &c., which do not need description, the following are noteworthy:—

521. Is the **ordinary V drill** for metal work.

522. **Flat point,** or "bottoming" drill.

523 & 524. **Countersinking drills** for metal.

525. **Centre bit** for wood.

526. **Twist bit** for wood; clears its own borings. There is a variety with rounded cutter edges.

527, 528, & 529. **Rock drills,** or "jumpers."

530. **Earth borer,** or mooring screw.

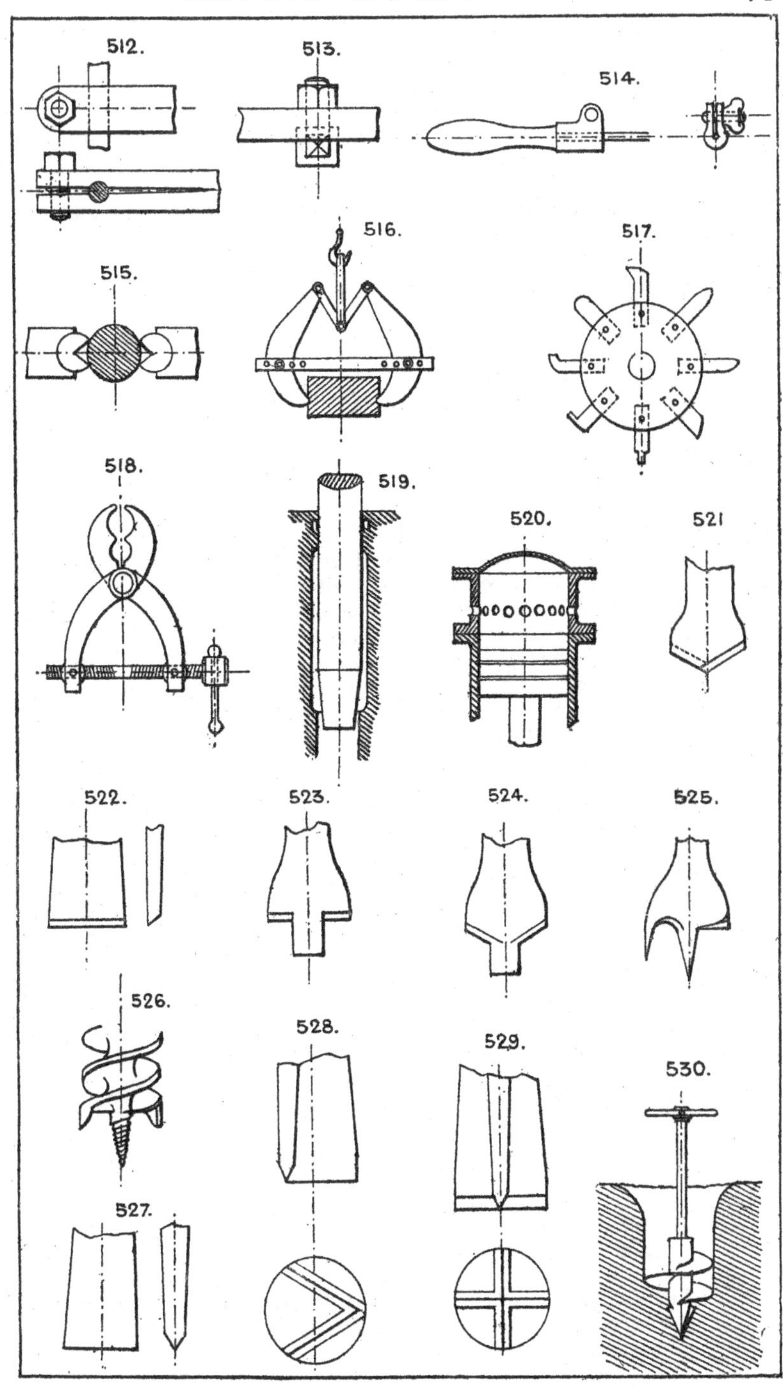
512.
513.
514.
515.
516.
517.
518.
519.
520.
521
522.
523.
524.
525.
526.
527.
528.
529.
530.

531. **Twist drill** for metal.

532 & 533. **Countersinking drills** for wood.

534. **Diamond drill** for rock; bores an annular hole, the core of which breaks out at intervals.

535 to 545. **Well boring tools** for different kinds of strata; tools for raising broken rods, &c.

546. **Hollow boring cutter** for cutting a shoulder on a central core; dowelling bit.

Section 31.—DIFFERENTIAL GEAR.

Devices to utilise the difference of velocity, or power, between two distinct moving parts.

547. **Equational box.** Two drivers A, A′, equally speeded in opposite directions, will drive the bevil gear at same velocity without revolving the spur gear; but any alteration in the relative speeds of A and A′, causes the bevil pinion to travel round, carrying the spur wheel C′ at a speed equal to half the difference of the two velocities. This gear is used on traction engines to drive the swivelling wheels round curves, where the proportionate velocities of the wheels will vary with the radius of the curve.

548. **Is a modification of 547.** The pinion A may be controlled in speed by any hand or automatic device, to vary the speed of the driven pinion B. The belt pulley C carries round the bevil wheel D, driving B at a speed varying with the motion given to A.

549. **Two wheels** (one of which has a different number of teeth to the other) gearing into one pinion; used for counters and slow motions of all kinds.

550. **Is an application of No. 549** by internal or epicycloidal gear to pulley blocks. Moore's patent (No. 1545) and Pickering's patent are examples. The arm shown is not required where two internal wheels are used with different numbers of teeth, but may be fixed to the pinion so that it is prevented from revolving, but retains its circular swaying motion; in this case, one wheel only is required, the speed being equal to the difference in number of teeth of the wheel and pinion at each revolution of the eccentric shaft.

551. **Weston's differential pulley block,** consisting of a two-grooved pitched chain-sheave having different numbers of teeth, in combination with a return block.

552. **Differential screws.** These may be both of the same hand, or one right and one left-handed, and any fractional speed secured by proportioning the pitches.

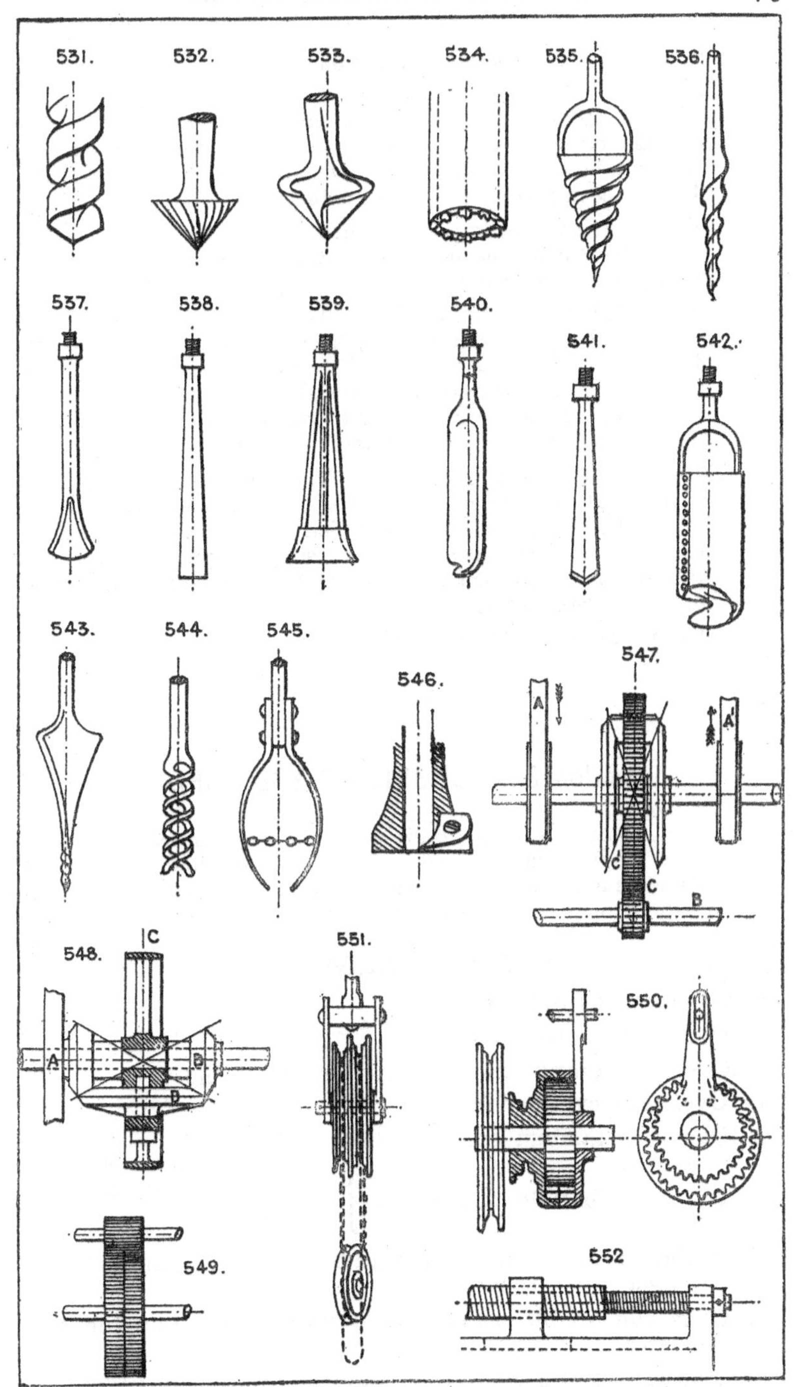
531.
532.
533.
534.
535.
536.
537.
538.
539.
540.
541.
542.
543.
544.
545.
546.
547.
A
A'
C
C
B
548.
C
A
B
D
549.
550.
551.
552

553. **Two-speed gear,** operated by a double clutch, which throws either pair into gear as required.

554. **Stewarts' differential gear.** Two cranks, one fixed to a sleeve and the other to a centre shaft, are driven round at varying velocities by a slotted crosshead revolving with the driving shaft. The two centres are not in the same line.

555. **Differential hydraulic accumulator.** The effective area of the ram is the annular shoulder, or the difference between the areas of the top and the bottom rams.

556. **Differential governing device.** The motive power drives A which winds up the large weight; the small weight tending to run down, drives the fan regulator, and the two weights are so adjusted that when the proper speed is attained, both weights are stationary; any change of speed causes them to run up or down, so actuating the regulation by the bell crank lever and rod.

557. **Varying differential regulator.** The upper rod A is connected to the regulator valve or other device, and it is capable of receiving motion from either the piston, which acts against a spring, or from the rod B attached to some positive reciprocating part, so that the nett movement of A is due to the difference of motion of B and the piston C.

Differential Worm Gear, No. 1559.

Section 32.—ENGINES (TYPES OF).

The following sketches are type drawings of the most important forms of Steam Engines in use, and are intended to afford a choice of outline arrangements from which in any scheme under consideration a selection may be made as a basis, without reference to details.

VERTICAL ENGINES.

558. **Overhead cylinder engine.**

559. **Overhead crank engine.**

560. **Overhead crank engine, cylinder oscillating.**

561. **Overhead cylinder engine,** with oscillating cylinder.

562. **Overhead tandem compound engine.**

563. **Overhead double compound engine.**

These can, of course, be reversed and the crank fixed overhead.

564. **Overhead one-crank compound oscillating engine,** cylinders at right angles and receiver between.

565. **Overhead double-crank compound oscillating engine,** with receiver.

566. **Overhead crank compound tandem oscillating engine.**

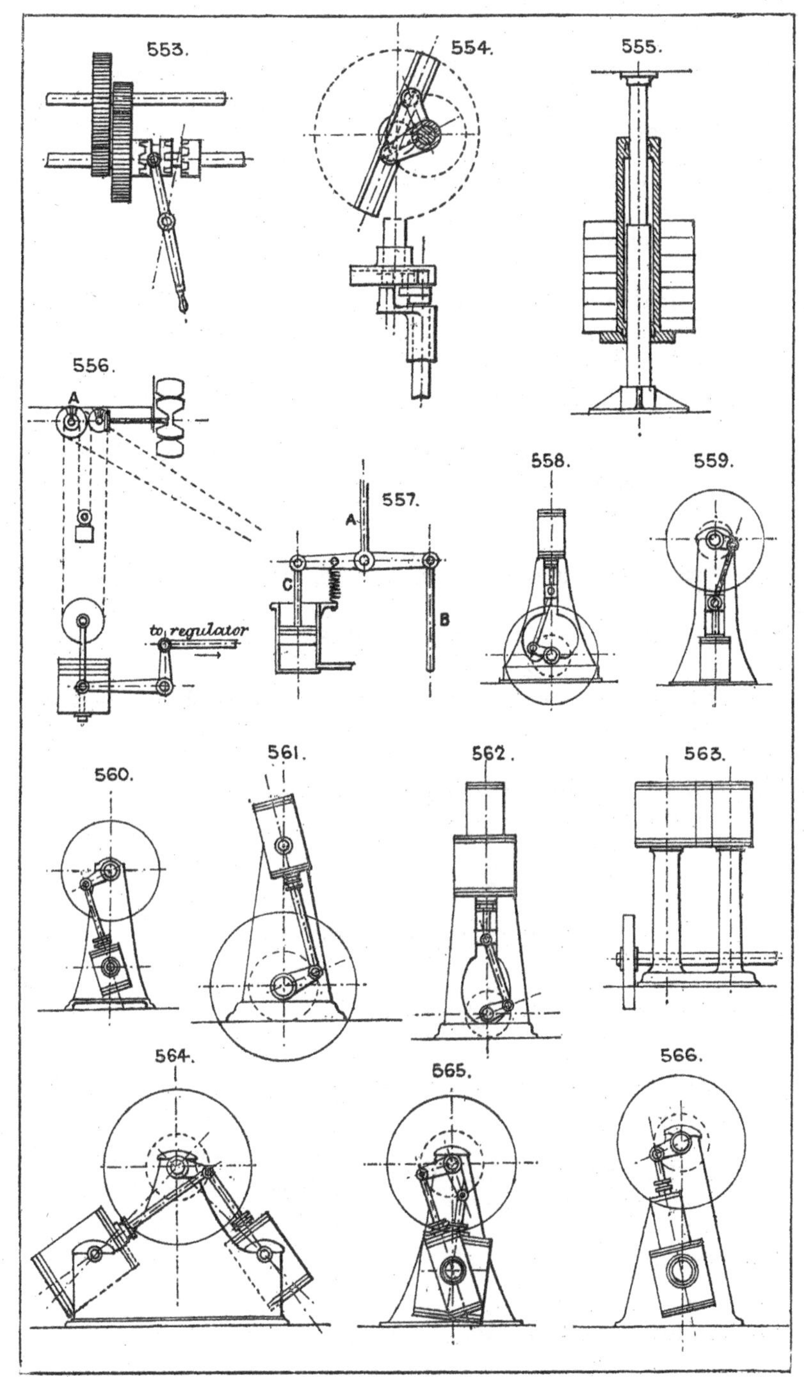
553.
554.
555.
556.
A
to regulator
557.
A
C
B
558.
559.
560.
561.
562.
563.
564.
565.
566.

567. **Vertical engine,** with top guides and double connecting rods.

568. **Vertical trunk engine.**

Note that the trunk plan is applicable to any of the preceding arrangements, and is employed where a very short engine is required.

569. **Vertical triple compound engine,** single acting cylinders. The high pressure steam acts first on the under side of the small piston, is then expanded into the annular under side of large piston, and finally expanded into the upper side of large piston.

570. **Vertical compound engine,** with annular cylinder. The central cylinder is the high pressure, and the annular cylinder the low pressure.

571. **Vertical annular cylinder engine,** with crank below.

572. **Vertical slotted crosshead engine.**

573. **Standard vertical engine;** a type largely used and possessing many good points.

574. **Double cylinder engine,** with T connecting rod. (Bernay's patent.)

HORIZONTAL ENGINES.

575. **Box bed engine,** high pressure.

576. **Double box bed engine,** coupled end to end.

577. **Oscillating cylinder engine,** with crosshead guides.

578. **Trunk engine.**

579. **Return crank engine.**

These can, of course, be duplicated side by side.

580. **Diagonal engine.**

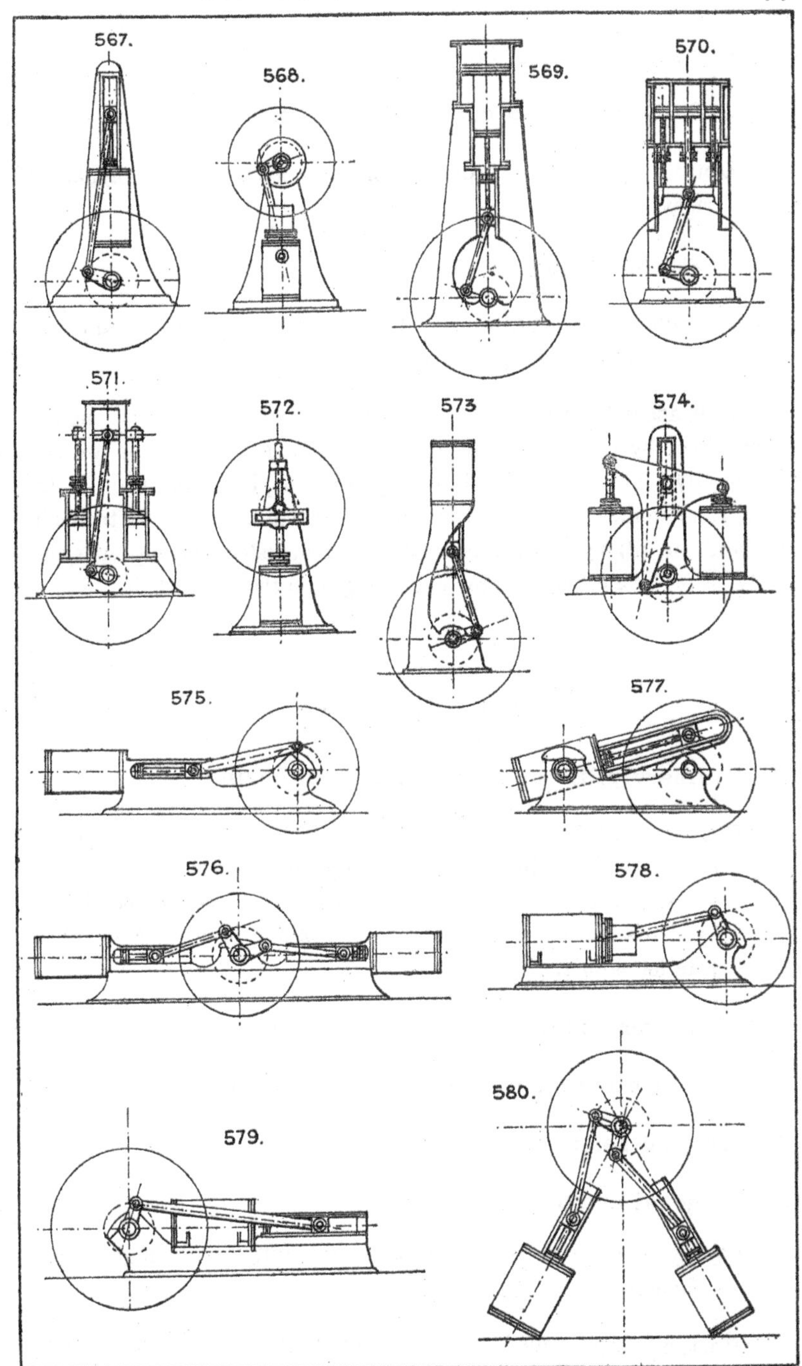
567.
568.
569.
570.
571.
572.
573
574.
575.
577.
576.
578.
580.
579.

581. **Horizontal tandem compound engine.**

582. **Galloway's oblique compound engine.**

583. **Double cylinder compound engine,** with receiver, cranks at right angles.

584. **Trunk bed engine.**

585. **Double piston engine.** The pistons are sometimes coupled to two crank pins at right angles.

Condensers (see Section 25) may be driven (*a*) from horizontal engines either direct by continuation of the piston rod, or (*b*) may be worked horizontally below by a vertical rocking beam coupled to the main crosshead, or (*c*) worked vertically below by a bell crank coupled to the main crosshead, or (*d*) by a separate small steam cylinder working independently, or (*e*) by a connecting rod or gearing from the crank shaft.

BEAM ENGINES, &c.

586. **Ordinary pillar and overhead beam engine.**

587. Has extended beam and double cylinders, either as a compound engine or one cylinder may form a pump or blast cylinder. In some designs the high and low pressure cylinders are placed side by side and coupled to the same end of the beam by a modified parallel motion.

588. **Side lever engine.**

589. **Plan of beam engine,** with compound cylinders.

590. **Walking beam engine.**

591. **Diagonal engine.**

592. **Three or four cylinder high-speed engine,** with single-acting cylinders.

593. **Vertical high-speed single-acting engine,** with one or more cylinders.

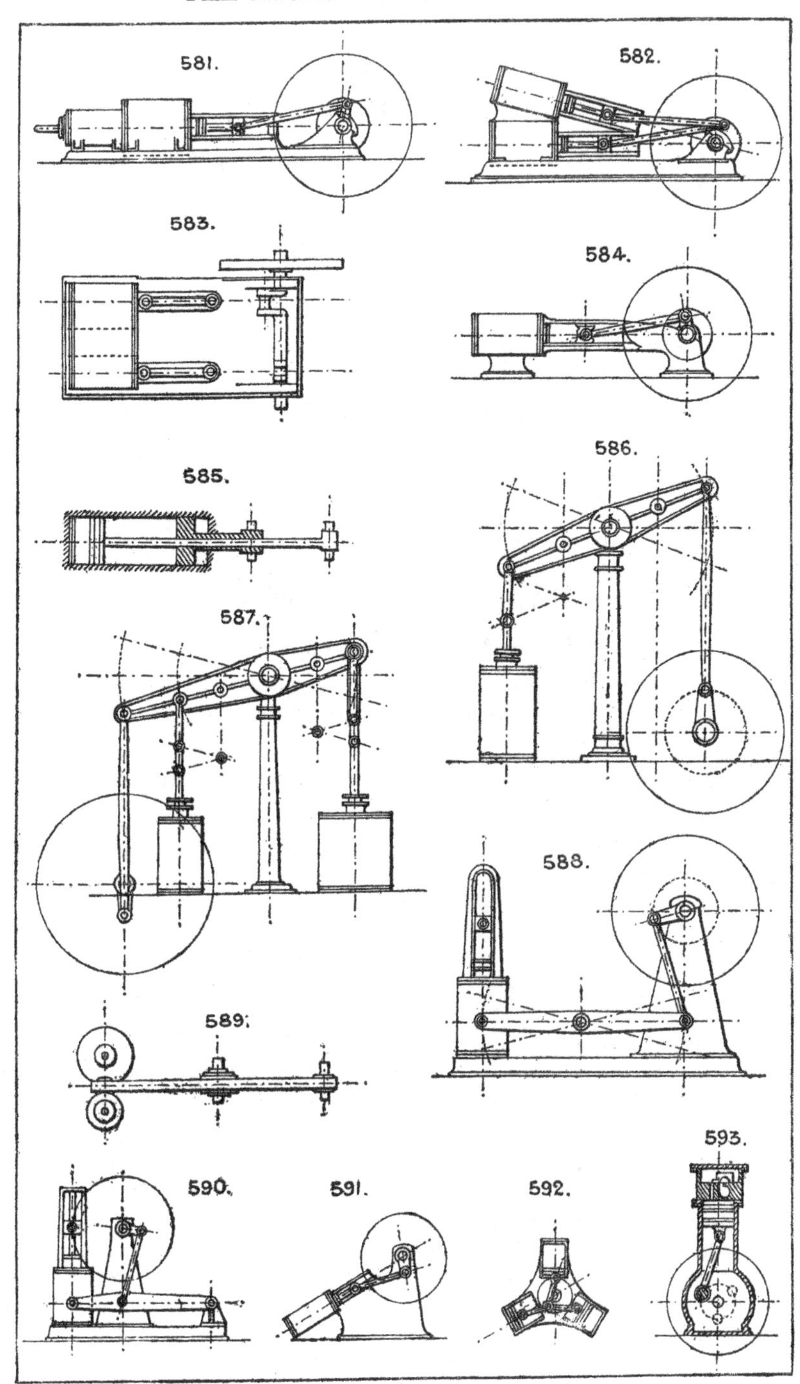
581.
582.
583.
584.
585.
586.
587.
588.
589.
590.
591.
592.
593.

Section 33.—ENGINES AND BOILERS COMBINED (see also Boilers, Sec. 6).

VERTICAL ARRANGEMENTS.

594. **In this engine the boiler forms the standard for support** of engine parts, but it is better to fix these on a vertical bed-plate bolted to the boiler, or as No. 595.

595. **Any type of vertical engine and any type of vertical boiler** can be combined on this plan.

596. **Vertical boiler** (any type) and horizontal engine (any type).

597. **Any type of vertical boiler,** with short horizontal engine on crown.

598. **Vertical boiler,** with cylinder sunk in the centre of crown.

599. **Overhead crank engine and boiler,** the latter forming the base to which the engine parts are fixed.

HORIZONTAL ENGINES.

600. **Loco.-type semi-fixed horizontal engine.**

601. **Loco.-type semi-fixed horizontal engine,** with engine on top. When placed on wheels this type constitutes the well-known "Portable."

602. **Horizontal semi-fixed boiler,** with circular shell and engine on top. (See No. 72.)

603. **Horizontal semi-fixed boiler,** with underneath fire-box. (See No. 71.)

Section 34.—ELLIPTICAL MOTION.

604. **Ellipsograph**; by gearing; the bevil wheel A is fixed, the other three revolving with the whole machine on the fixed central standard, the distance A′ should equal the difference between the major and minor axes of the ellipse.

605. **Performs the same operation** in a similar way; A is fixed; B is same diameter as A; and C = ½ diameter of A and B.

606. **Common trammel or ellipsograph.**

There are other forms of apparatus for *drawing* ellipses merely (see Knight's Dictionary of Mechanics). See Trammel Gear, Sec. 40.

607 & 608. **Two forms of ellipsographs** or elliptical cranks.

See also No. 144.

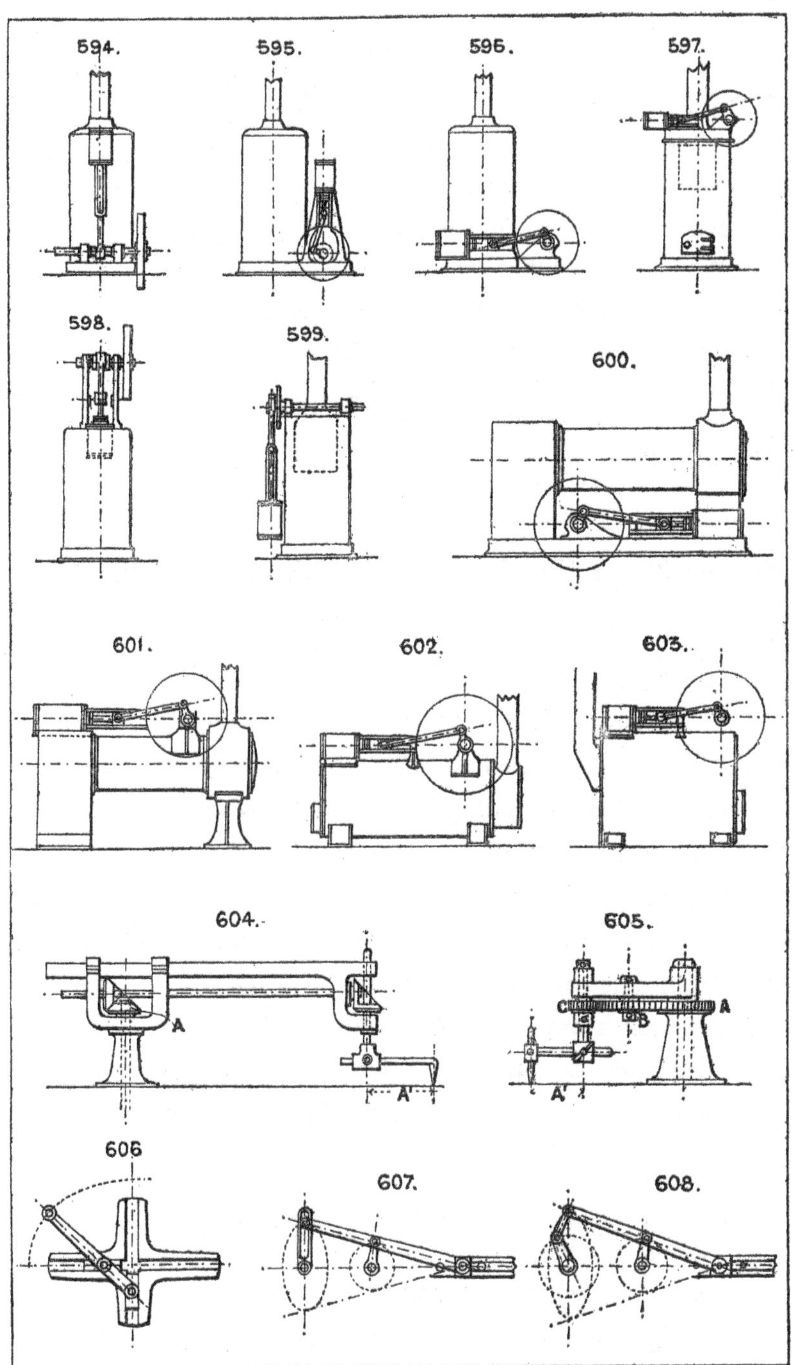
594.
595.
596.
597.
598.
599.
600.
601.
602.
603.
604.
A
A'
605.
C
A
B
A'
606
607.
608.

Section 35.—ELASTIC WHEELS.

Wheels with rubber tyres are the common form. Wheels with a loose flat tyre outside a fixed tyre and with various forms of springs inserted between the tyres. (See Springs, Section 80.)

Some have adopted bent steel spokes to obtain sufficient elasticity.

609. **Huxley's wheel,** with spring tyre and jointed spokes.

610. **Wheel with double tyres** and intermediate springs.

611. **Two plans of bent spoke, spring formation.**

612. **Two plans of bent spoke, spring formation.**

613. **Has an outer elastic tyre** and an inner rigid ring, to which the tension or compression springs are fixed.

614 & 615. **Sections of rubber tyres.**

Section 36.—EXPANDING AND CONTRACTING DEVICES.

Common expedients for these purposes are—jointed folding rods; the telescope tube; net work; diagonally crossed and jointed bars; springs (see Section 80); lazy tongs (No. 623).

616 **Telescopic ram hydraulic lift.** (See also No. 1217.)

617. **Parallel bar expanding grille,** or gate.

618. **Parallel bar expanding grille,** with lazy tongs motion.

619. **Modification of 618.**

620. **Venetian blind.**

621. **Venetian blind,** but without revolving motion to the laths or slats.

622. **Perforated bar and hooked rod suspender.**

623. **Lazy tongs expanding connecting rod.**

624. **Four-guide expanding link device,** for varying motion.

625. **Thorburn's tube expander,** operated by a cone and ring of conical rollers.

626. **Gasometer.**

627. **Timms' expanding boring tool.** Operated by a central cone and three or more diagonal feathers.

Expanding Mandrel, No. 507.

Expanding Chucks, Nos. 489, 491, & 506.

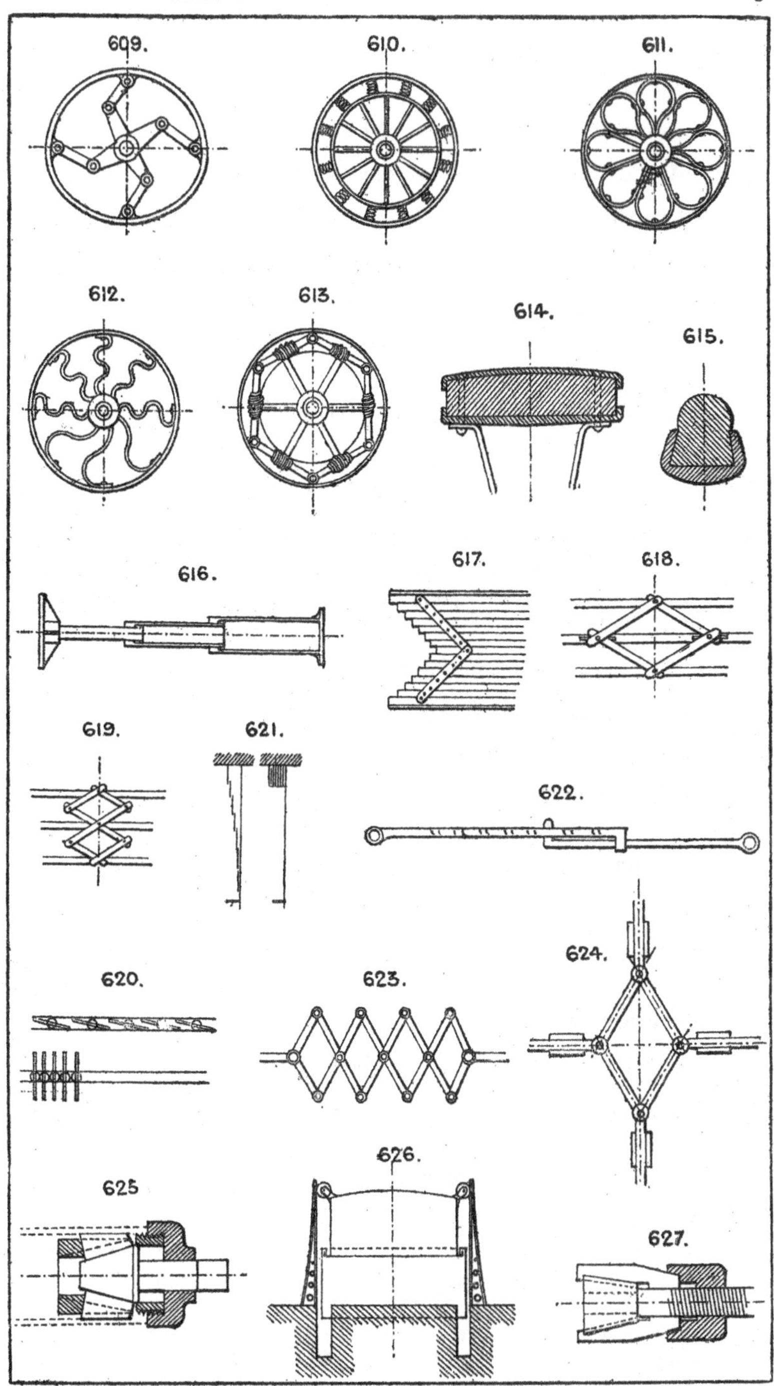
609.
610.
611.
612.
613.
614.
615.
616.
617.
618.
619.
621.
622.
620.
623.
624.
625
626.
627.

628. **Expanding basket.**

629. **Expanding socket.**

630. **Expanding grating,** formed of bent steel laths on edge.

631. **Bridge or flap** between cars, having buffers.

632. **Expanding core barrel,** in three parts, expanded by a wedge.

633. **Expanding mandril or chuck.** See also Section 28.

Section 37.—FASTENING WHEELS TO SHAFTS.

Besides the ordinary plan of shrinking them on while hot, the following are the chief devices in use:—

634. **Square shaft and single key.**

635. **Square shaft and two keys** at right angles; two keys should always be used for a square shaft, unless it has been machined to fit to the hole.

636. **Round shaft and hollow key.**

637. **Round shaft and flat key.**

638. **Round shaft and sunk key.**

639. **Staked fastening,** four keys, usually on flats cut on shaft, but better if slightly sunk into shaft.

640. **Set screw.** Cannot be depended on for any but light strains.

641. **Taper pin.**

642. **Split pin**; always used where a pin, bolt, or centre is liable to work loose.

643. **Cotter and slot.**

644. **Screwed pin** through shaft and boss of wheel.

645. **Octagonal shaft of cast iron,** with four keys, or the four keys may be cast on shaft.

646. **Cotter through side of shaft.**

647. **Large wheels** are sometimes wedged with iron and wood wedges all round on a square or octagonal shaft having feathers cast on it.

648. **Set screw,** tapped half into shaft and half into wheel.

649. **Screwed shaft nut and clamping plates**; used for emery wheels, grindstones, circular saws, and milling cutters.

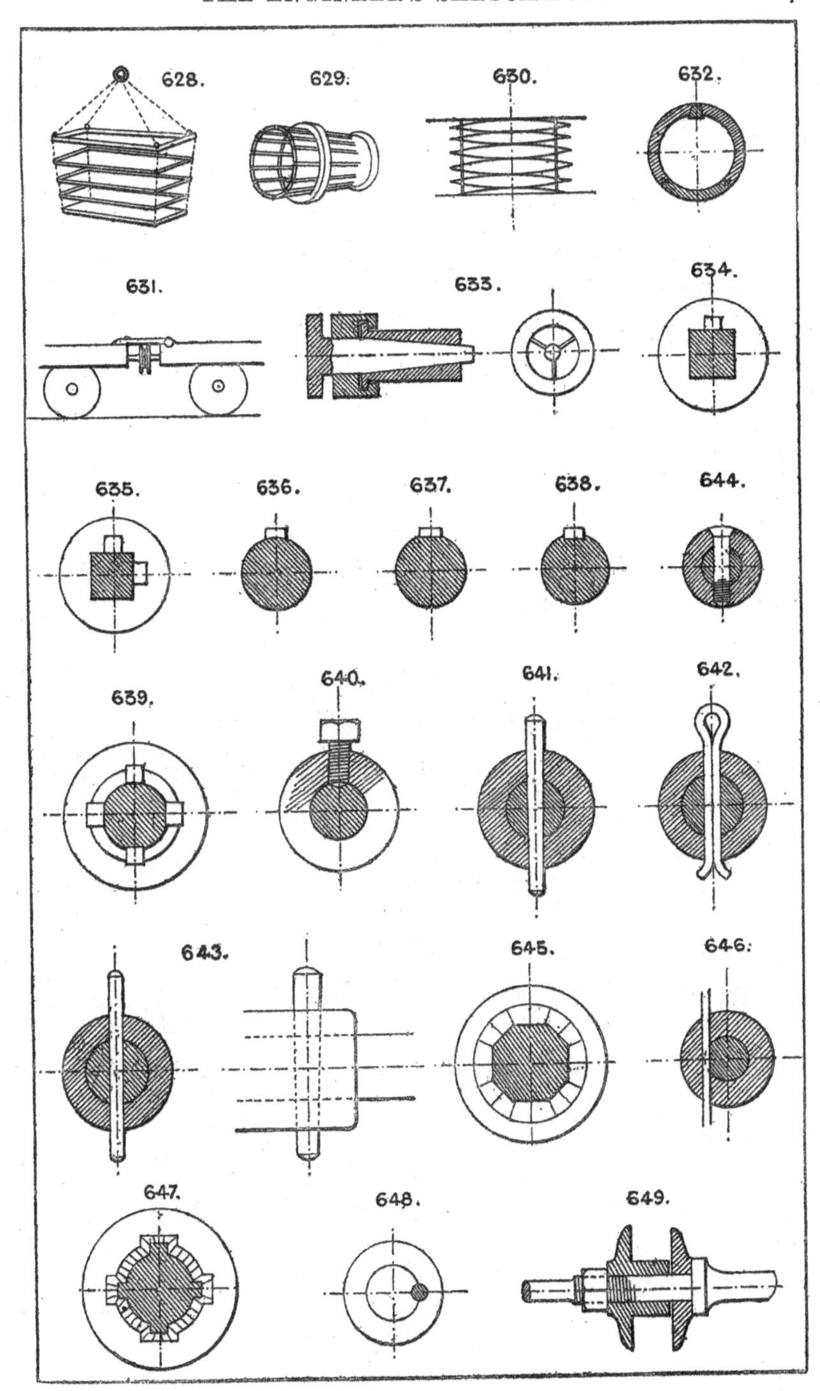
628.
629.
630.
632.
631.
633.
634.
635.
636.
637.
638.
644.
639.
640.
641.
642.
643.
645.
646.
647.
648.
649.

650. **Screwed end and nut,** the hole in wheel being square, or round and fitted with key.

651. **Gib head taper key.**

652. **Plain taper key.**

653. **Taper round pin.**

654, 655. **Split pins** (round).

656. **Cotter and split pin.**

657. **Cotter and nut.**

658. **Dovetail taper key,** or fixing for projection or bracket.

659. **Self-locking pin**; cannot work out.

660. **Split collar and ring fastening,** sometimes used instead of a nut and screwed end; the inner ring is in halves.

661. **Piston rod fastening.**

662. **Locking feather and wedge** fastening, prevents end motion.

663. **Railway chair key.**

Section 38.—FRICTION GEAR.

Various forms of friction gearing are much used, the chief objection to this kind of gear being the excess of pressure on the bearings required to give sufficient grip to drive the gear.

664. **The common form of flat-faced friction gear** for hoisting purposes, &c. See No. 1211. The required pressure is given by a weighted lever.

665. **Friction bevils,** plain faces, for governor driving, &c.

666. **Friction bevils,** the pinion being usually of hard leather; the pressure may be applied in the direction of either of the arrows.

667. **Multiple V gear.** A common mistake is to run these too deep in gear; the narrower the surfaces in contact, short of the seizing or crushing point, the less the power wasted in friction.

668. **A very small pinion of leather, wood, or rubber** is frequently driven by a large driving wheel for obtaining high speed with steadiness, for driving dynamos, fans, &c.

669. **Disc wheel and rubber pinion,** arranged to reverse motion or vary speed. See No. 1595. The motion is reversed by throwing either wheel into gear with the pinion, and the speed varied at will by raising or lowering the pinion.

670. **Wedge friction gear.**

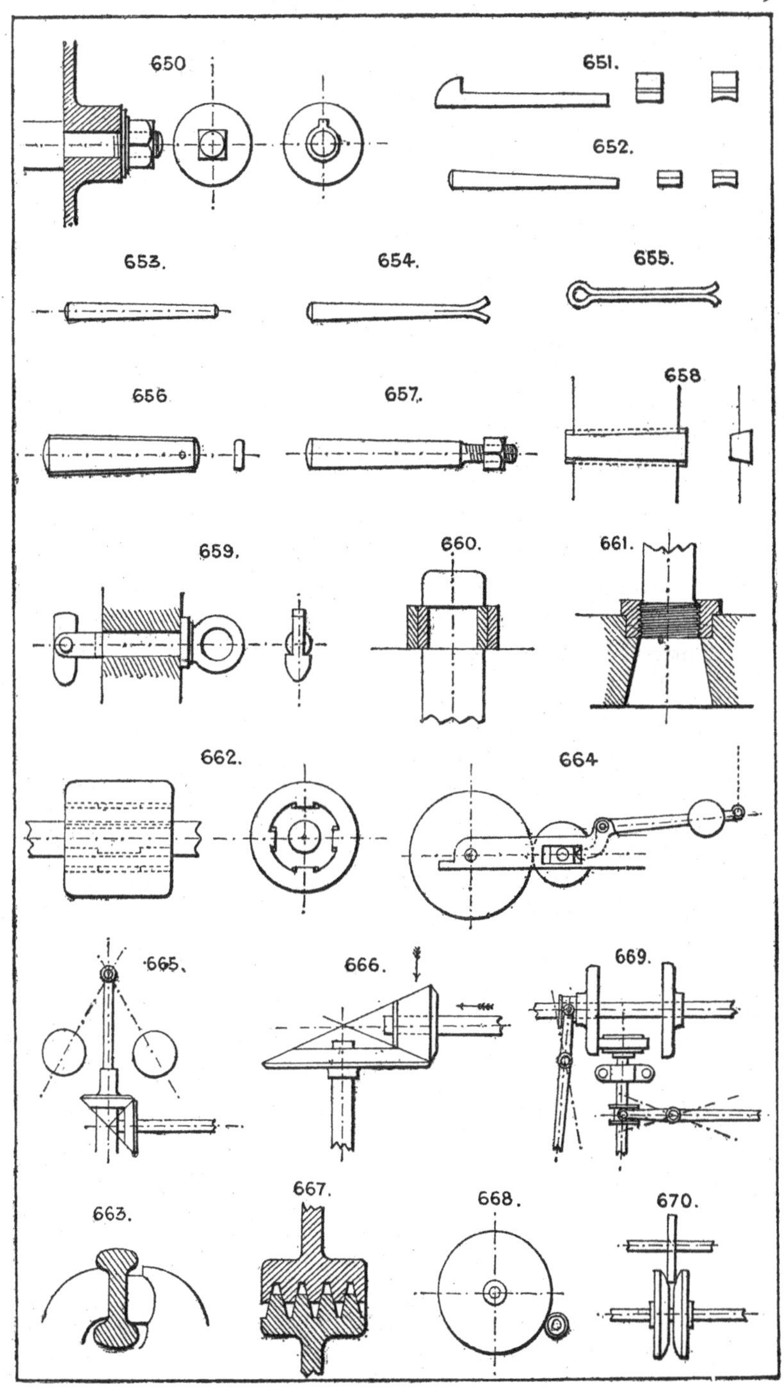
650
651.
652.
653.
654.
655.
656
657.
658
659.
660.
661.
662.
664
665.
666.
669.
663.
667.
668.
670.

671. **Coupled bearings** for friction gear, to allow of any required pressure or "bite," the strains being self-contained.

See also No. 737, 738, 1294.

Section 39.—GUIDES, SLIDES, &c.

PISTON-ROD GUIDES.

672. **Two bars and crosshead**; these must be far enough apart to allow for the angle of the connecting rod.

673. **Four bars, crosshead, and slide blocks**; the connecting rod working between the two pairs of guides.

674. **Bar and slipper.**

675. **Adjustable slipper**; there are other adjustments for wear by wedge pieces, similar to No. 19. See also No. 21.

A plain guide bush is sometimes used as No. 682, and a forked connecting rod with long fork coupled to the gudgeon or crosshead.

676. **Section of No. 673 and alternative crosshead** for two round bar guides.

677. **Slide bed and slipper.**

678. **Section of trunk guide**, cast with engine bed and bored out.

679. **Oscillating cylinder piston head guides.**

680. **Oscillating fulcrum** in lieu of guides.

681. **Diagonal crosshead and guide bars**, to allow the crank and connecting rod end to pass the guide bars.

VALVE ROD GUIDES.

682, 683, 684, & 685.

GUIDE ROLLERS.

686 & 687. **Guide rollers** for ropes, &c.

688 & 689. **Guide rollers** for bars of various sections.

LIFT AND HOIST GUIDES.

690. **Cage guided by four corner posts.**

691 & 692. **Cage runs on two vertical rails**, and is steadied by a third guide. For large cages. Small cages only require guides on one side, as 692.

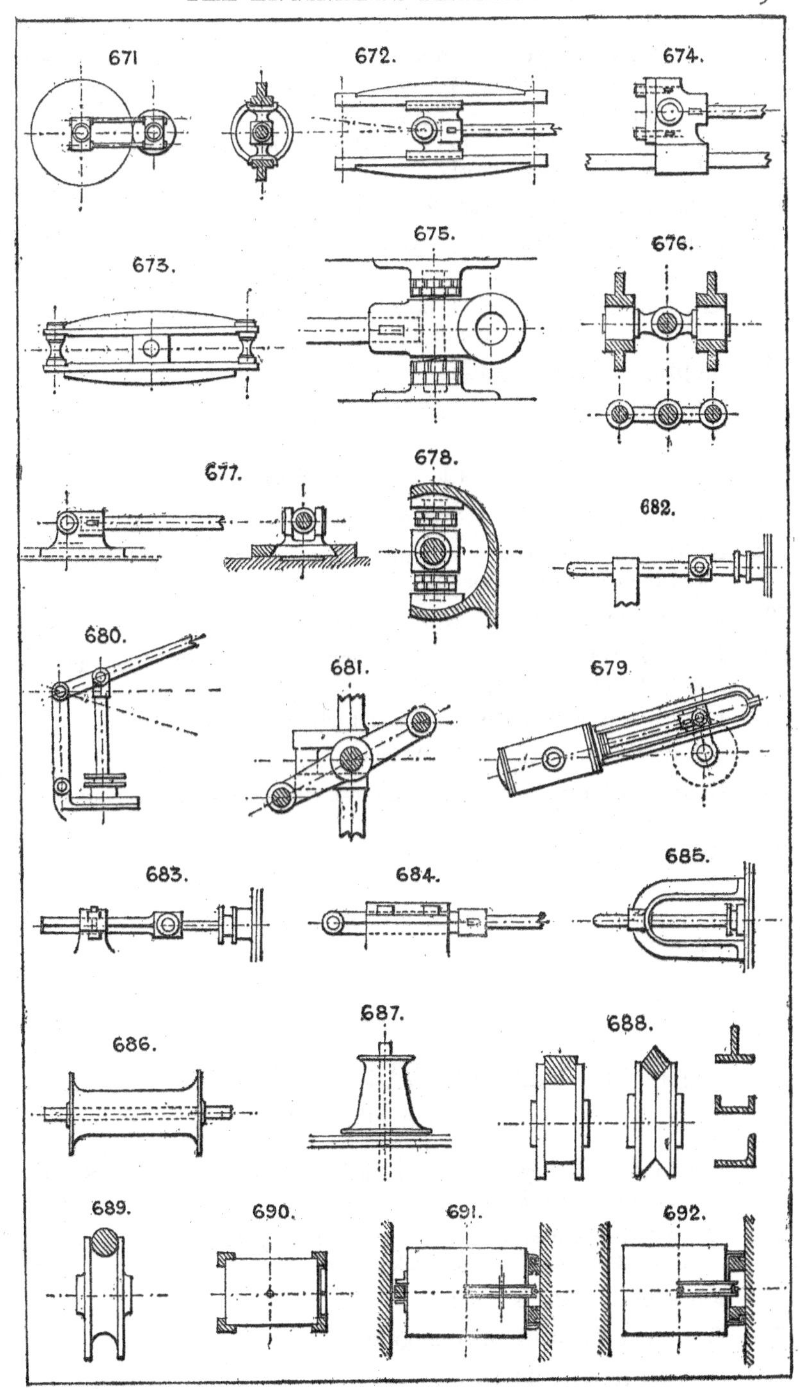
671
672.
674.
673.
675.
676.
677.
678.
682.
680.
681.
679
683.
684.
685.
686.
687.
688.
689.
690.
691.
692.

693. **Iron wire or rod guides,** strained tight, are sometimes used, especially in mines, as guides for the cage.

694. **Planished round iron guides,** with half round fixing brackets and runners attached to cage; these guides are equal to planed bars and much less costly.

695. **T L or ⊔ iron guides,** for goods lifts.

696. **Wire rope guides,** with separate pair of wood guides for balance weight.

697. **Intermediate guides** for double cage lifts; for large cages extra guides at each side should be used.

698. **Sloping carriage guides.**

699. **Vertical bracket cage guides.**

MACHINE GUIDE BEDS.

700. **Double V bed,** with set screw adjustments.

701. **Guide bed** for planing machine, or any machine where the bed is not liable to lift in working.

702. **Round bar and flat guide bed.**

703. **Deep V guide;** much used for crossheads, tool boxes, &c., requiring accurate movement.

704. **Lathe bed** with square guides and adjustments for wear.

705. **Planing machine,** double V bed.

706. **Crosshead** for two single bar guides, with renewable wearing strips and square guide surfaces.

707. **Radial slide** for tool box.

ROPE GUIDES.

708, 709, & 710.

Section 40.—GEARING, VARIOUS DEVICES IN (not otherwise classed).

711. **Conical rotatory gear.** Applied to reaping machines. See also Pan Screen, No. 1264.

712. **Triangular eccentric,** used to obtain a pause of one-third revolution at each end of the stroke.

713. **Face plate worm gear.**

714. **Double rack and pinion gear.**

715. **Double gear wheels.**

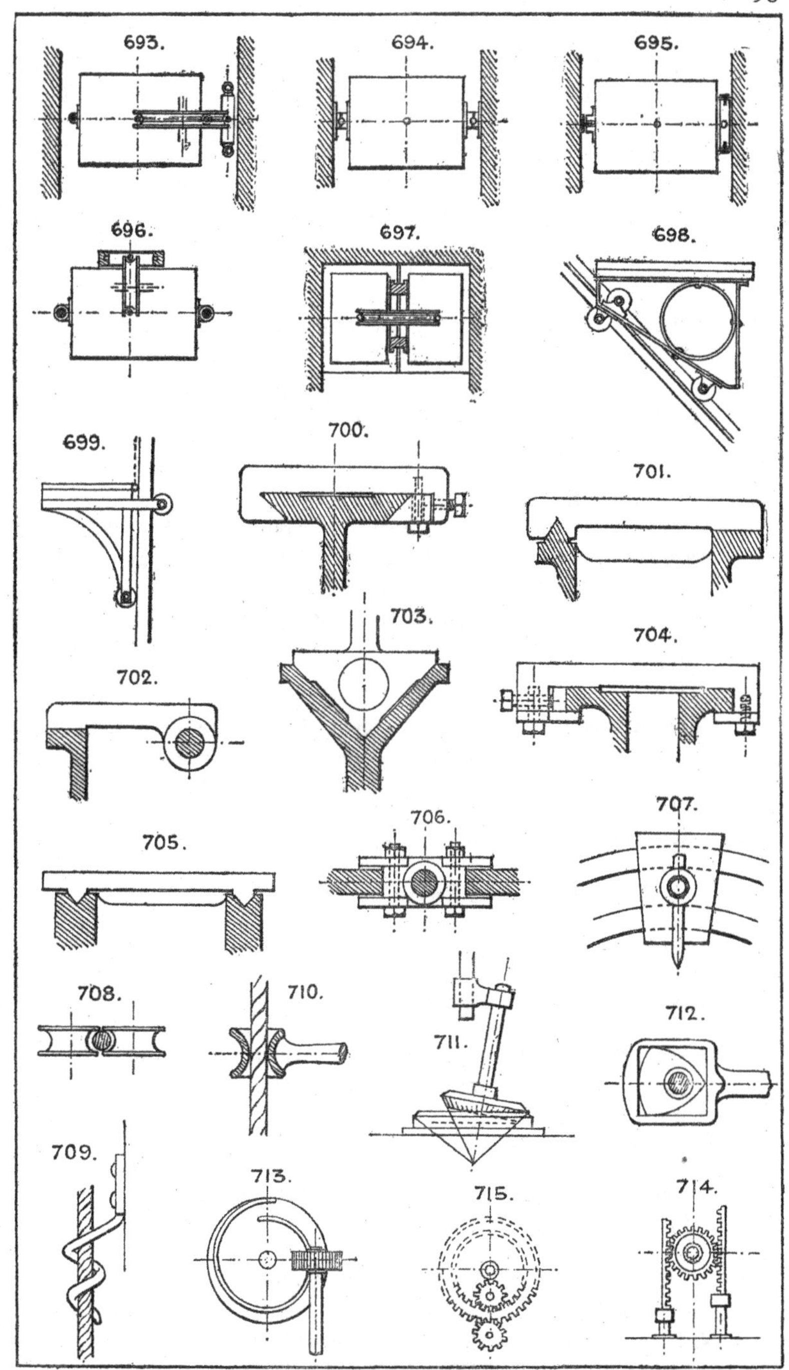
693.
694.
695.
696.
697.
698.
699.
700.
701.
702.
703.
704.
705.
706.
707.
708.
710.
711.
712.
709.
713.
715.
714.

716. **Eccentric gearing**; the wheel A being fixed on a crank pin in the driving wheel B, drives the dotted gear at a speed proportionate to the diameters of the wheels A and the driven wheel.

717 & 718. **Forms of epicyclic or planet gear.** Several modes of driving these may be employed by fixing one or other of the three wheels, the other two revolving. See Differential Gear, Section 31.

719. **Multiple trammel gear.** The pinion is half the diameter of the wheel, and makes two revolutions to one of the wheel.

720. **Trammel crank gear**; the crank revolves once to two double strokes of the rod.

721. **Knight's noiseless gearing**, for two shafts running in opposite directions. Each shaft has two equal cranks at right angles, which are coupled by links to rocking arms, which are also coupled in pairs.

722. **Eccentric variable speed toothed gear.**

723. **Scroll bevil gear.**

724. **Segment reversing gear**, to obtain two speeds in portions of one revolution, and in opposite directions. See Reversing Gear, Section 74.

725. **Snail wheel**, or scroll ratchet.

726. **Combined spur and bevil wheel.**

727. **Double screw gear**, for steering gear, &c.

728. **Angular ball-jointed crank motion.**

729. **Crank gearing** between two shafts running in the same direction. See No. 187. The cranks should be similar to Nos. 174 or 175.

730. **Snail worm gear.**

731. **Diagonal engine or pump**, with bevil gear revolving motion and three or more cylinders.

732. **Angle coupling** on Dr. Hooke's principle. See No. 292.

733. **Worm and crown gear.** Used in chaff machines; useful to obtain a slow feed on two shafts in opposite directions.

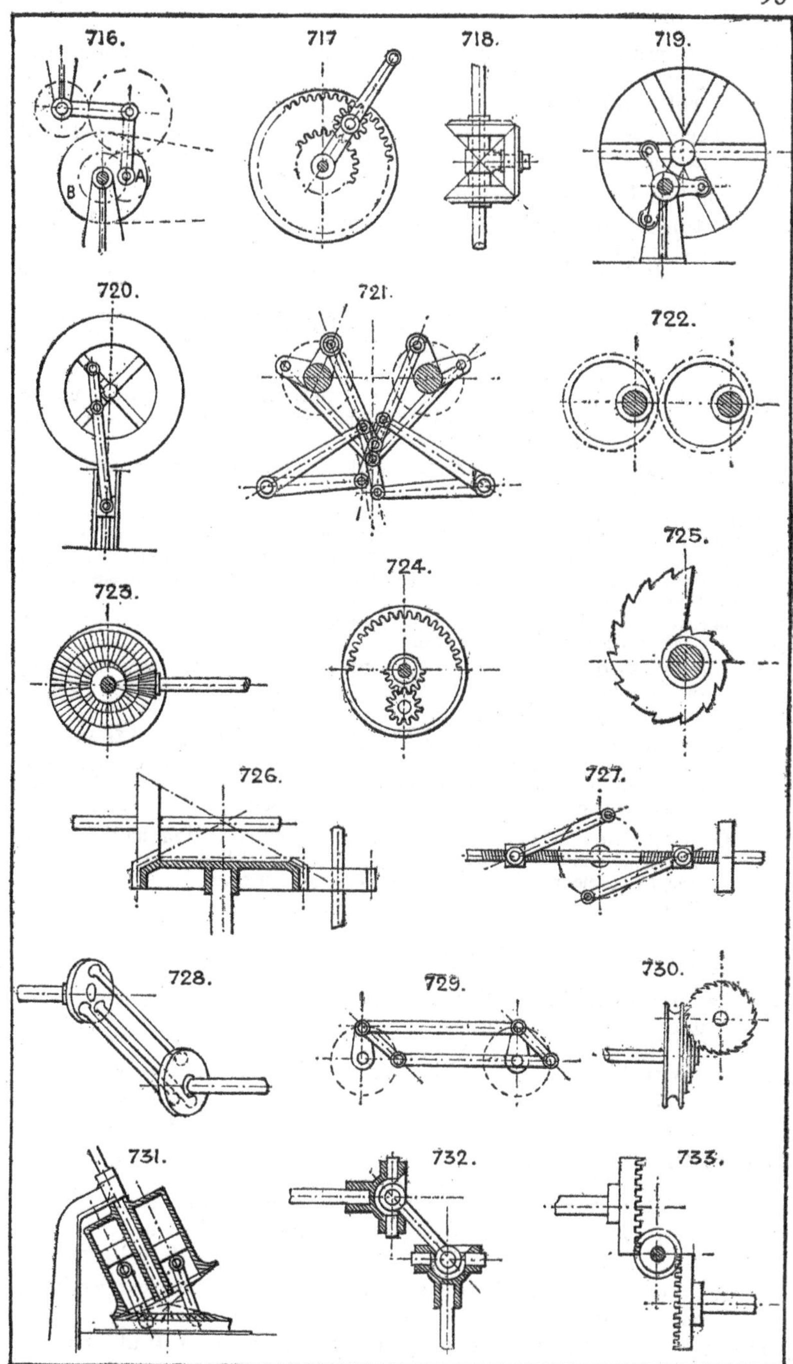
716.
A
B
717
718.
719.
720.
721.
722.
723.
724.
725.
726.
727.
728.
729.
730.
731.
732.
733.

734. **Ball wheel,** with limited angular traverse gearing into one or two pinions.

735. **Scroll and rack.**

736. **Variable speed gear,** from an elliptical or other irregular-shaped driving wheel, combined with a tied idle intermediate wheel.

737. **Spring friction grip wheels.**

738. **Intermittent reversible feed motion.** The pinion is of leather, and drives the segment till it runs out of gear; when the machine is reversed it travels an equal distance the opposite way.

Section 41.—GOVERNING AND REGULATING SPEED, POWER, &c.

739. **Is a device for varying the opening of a main valve** (connected to rod A) by the pressure on the small piston, which moves it against the tension of a spring.

740, 741, 742, & 743. **Types of centrifugal governors,** of which numerous varieties are in use.

Pumping engines may be governed by allowing the pressure of water in the rising main to accumulate in a stand pipe or equivalent device until it stops the engine by excess of pressure. To prevent such an engine running away a catch is used, kept open by the pressure of water; when the pressure falls below a certain point the catch is released and closes the throttle valve.

Steam engines may also be safeguarded in the same way by a catch which is released when the governor becomes fully expanded.

744. **Screw and nut device,** to control the travel of any machine, such as a lift, by reversing the belt or throwing out a catch after any specified number of revolutions.

745. **The cataract** is one of the oldest governing devices. It consists essentially of a vessel which is filled with water by one stroke of the engine, and empties itself through an adjustable orifice during the return stroke, the valve motion being prevented from reversing till the water is all discharged.

746. **Gas engine governor.** Rod A has a reciprocating motion from the engine, and the spur on lever B strikes the end of the gas valve slide when brought in line with it by the motion of the governor, thus supplying gas only when the governor falls to a certain point.

Section 42.—HYDRAULIC MULTIPLYING GEAR.

747. **Is the ordinary "chain and sheave" multiplying gear,** unequally geared, thus—

Ram end.		Cylinder end.			
With 1 sheave	..	1 sheave	..	it is geared	3 to 1.
„ 2 „	..	2 „	..	„	5 to 1.
„ 3 „	..	3 „	..	„	7 to 1; &c.

748. **Is the same plan,** but equally geared—

Ram end.		Cylinder end.			
With 1 sheave	..	No sheaves	..	it is geared	2 to 1.
„ 2 „	..	1 sheave	..	„	4 to 1.
„ 3 „	..	2 „		„	6 to 1; &c.

749. **An arrangement of the sheaves** suitable for vertical working, geared 8 to 1.

750. **An arrangement of the sheaves** suitable for vertical working, but geared 6 to 1.

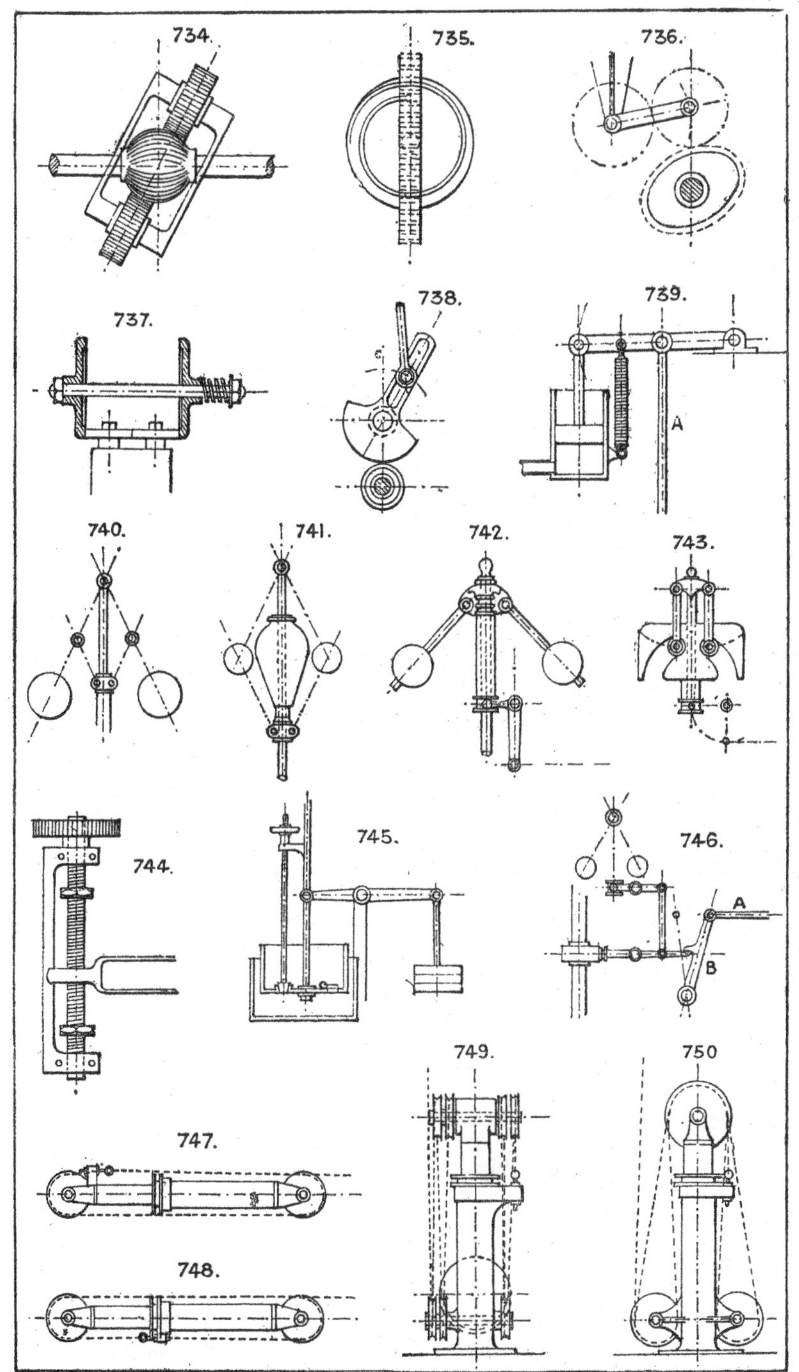
734.
735.
736.
737.
738.
739.
A
740.
741.
742.
743.
744.
745.
746.
A
B
747.
748.
749.
750

751. **An arrangement of the sheaves** suitable for vertical working, but geared 4 to 1.

752. **Stevens and Major's patent** for horizontal working. The angle of the chain helps to support the weight of the ram.

753. **Modification of 752,** sometimes used, and suitable for both horizontal and vertical positions, with any required multiplication of speed.

754. **Rack gear;** short stroke piston cylinder plan.

755. **Double rope vertical ram gear.**

756. **Arrangement with the sheaves** all at head of cylinder.

For Telescopic Hydraulic Lift, see Nos. 1217 & 616.

Section 43.—HOOKS, SWIVELS, &c.

For Chains and Links, see Section 11.

757. **Double or match hook.**

758. **Split link.** See also the common Key Ring.

759. **Self-locking hook,** with inclined shoulder and pin.

760. **The common "Lewis."**

761. **Self-gripping claw grab.** See also 516, 505.

762. **Grab bucket,** on same plan as last.

763, 764, & 765. **Double S links.**

766. **Hook with rope grip.**

767. **Snap hook.**

768. **Snap link.**

769. **Slip hook** for a monkey or pile engine; a rope is attached to the eye in end of lever which pulls the loop link away from the bottom link to which the "monkey" is suspended, allowing it to fall.

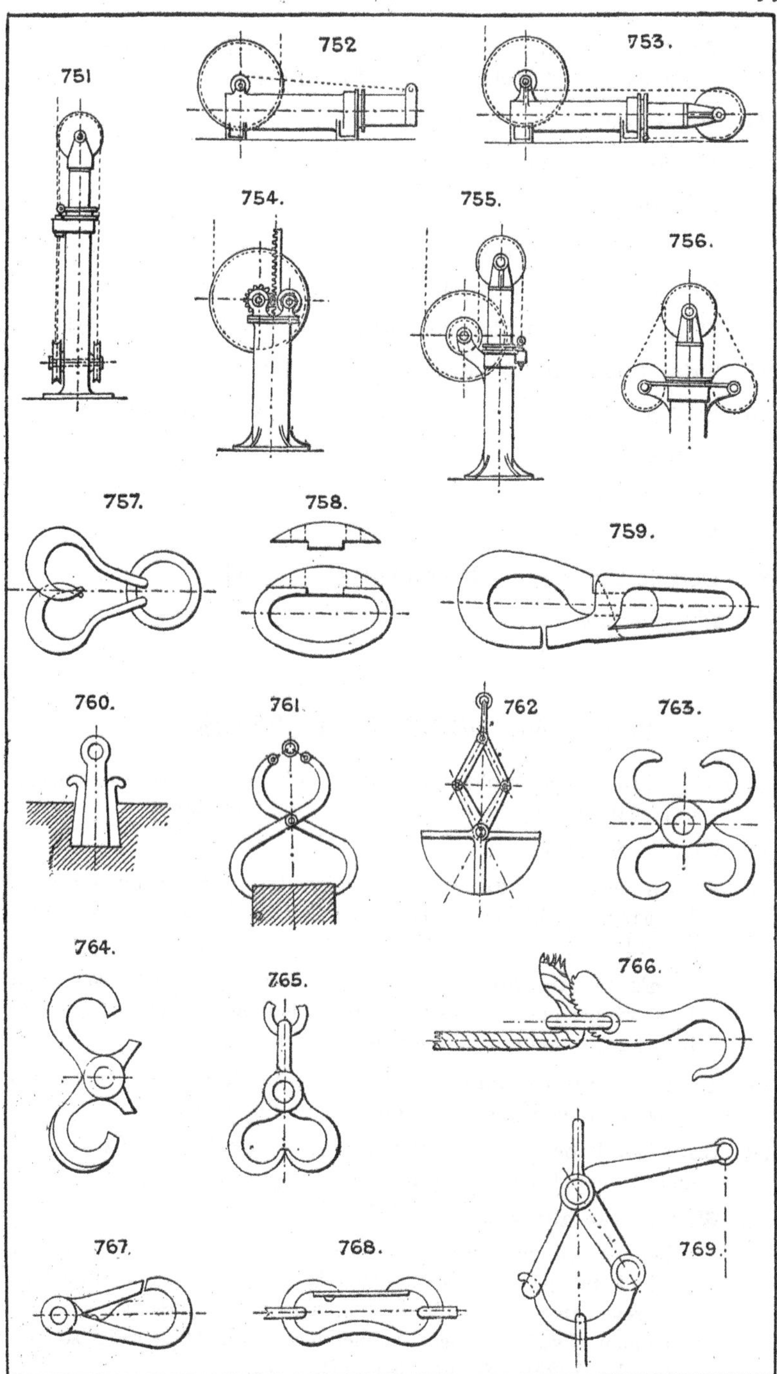
751
752
753.
754.
755.
756.
757.
758.
759.
760.
761
762
763.
764.
765.
766.
767.
768.
769.

770. **Automatic slip hook**; slips the T end of the "monkey" by the curved arms striking the sides of a fixed stop hole.

771. **Draw bar hook,** self-locking.

772. **Fixed bar hook,** with snap.

773. **Slip hook.**

774. **Hook,** with mousing ring.

775. **Crane hook,** with swivel.

776. **Double swivel links,** inserted in a chain to take out the twist.

777. **Triangular link,** to attach two chains to one.

778. **Safety link.** Has a flat on link to slip in notch of hook.

779. **S link.**

780. **Split link.**

781. **Bolt shackle.**

782. **Double link and bolt connection** for chain.

783. **Pin shackle.**

Section 44.—INDICATING SPEEDS, &c.

784 & 785. **Hand (portable) indicator,** to indicate speed of revolution of a shaft, &c., by simple wheel work and dial plate.

786. **Governor gauge.**

787. **Steam engine indicator,** of which there are many varieties. Macnaught's, Richards', Darke's, Kraft's, Casartelli's, &c., are examples.

788. **Morin's dynamometer.** Consists of two belt pulleys connected by a spring; one receives the strain of belt, and the other transmits it, the spring indicating the tension on the belts.

789. **Regnier's dynamometer** indicates the tension on the connections by contraction of the spring operating a dial plate.

790. **Bourdon tube pressure indicator.** The tube is of flat section, and its curved portion expands with the pressure.

791. **Worm gear and dial.**

Other forms of pressure gauges are—1st. The mercurial gauge, in which the pressure is indicated by the height of a column of mercury in a glass tube. 2nd. The water gauge, in which a column of water replaces the mercury. 3rd. The spring balance (see No. 1729). See also Nos. 1730, 1728.

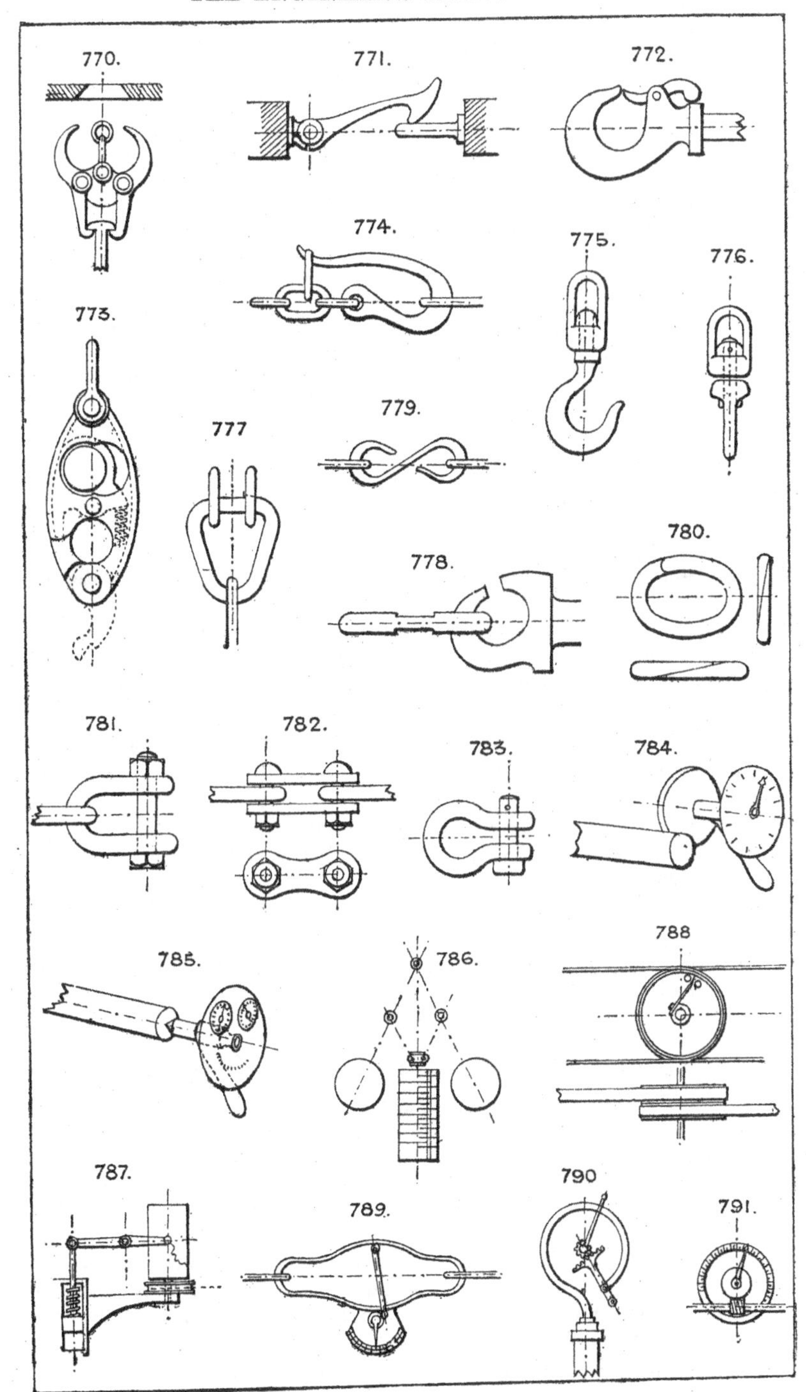
770.
771.
772.
773.
774.
775.
776.
777
778.
779.
780.
781.
782.
783.
784.
785.
786.
787.
788
789.
790
791.

792 & 793. **Winding engines** are provided with indicators on the principle of No. 744. The travelling nut has a pointer whose position on a vertical graduated scale shows the position of the cage in the pit.

Vertical scale indicators are also employed to show the level of water in tanks, reservoirs, &c. See No. 1730.

Water tube indicators are employed to show the level of water in boilers, &c., as also gauge cocks fixed at various heights in the boiler.

Section 45.—JETS, NOZZLES, AND INJECTORS.

794. **Straight jet,** for long distances.

795. **Short jet.**

796. **Rose jet,** for spreading.

797. **Fan jet,** or spreader.

798. **Blast tuyere.**

799. **Smith's tuyere and water bosh.**

800. **Jet aspirator,** for inducing a mixed current of air and water or steam.

801. **Steam jet pump**; the steam enters by the central jet and causes a vacuum, into which the water rises by the branch pipe.

802. **Insufflator** for steam and air blast; used also as a petroleum injector, &c.

803 & 805. **Spray jets**; the liquid rises by gravity at the small vertical nozzles, and is driven in a spray or mist by a cross blast of air from the horizontal jets.

804. **Injector.** The varieties of this contrivance are too numerous to specify. See Graham's, Gifford's, Hall's, Hancock's, and others in common use.

806. **Plain or spreading jet.** The eight vanes can be pushed into the jet of water to cut it up by moving the sliding ring.

807. **Ventilating jet or aspirator,** with several lateral openings for inducing a current.

Section 46.—JOURNALS, BEARINGS, PIVOTS, &c.

See also Section 70.

808. **Plain or solid pedestal.**

809. **Half bearing,** sometimes used without a cotter.

810. **Half bushed bearing,** having a half brass on the lower side only.

811. **Chambered long bearing.**

812. **The ordinary double brassed pedestal** or plummer block; sometimes made with the cap and joint of brasses at an angle of 45° when the shaft is subject to horizontal thrust. Numerous modifications of this bearing exist.

813. **Slot bearing** for rising and falling spindle.

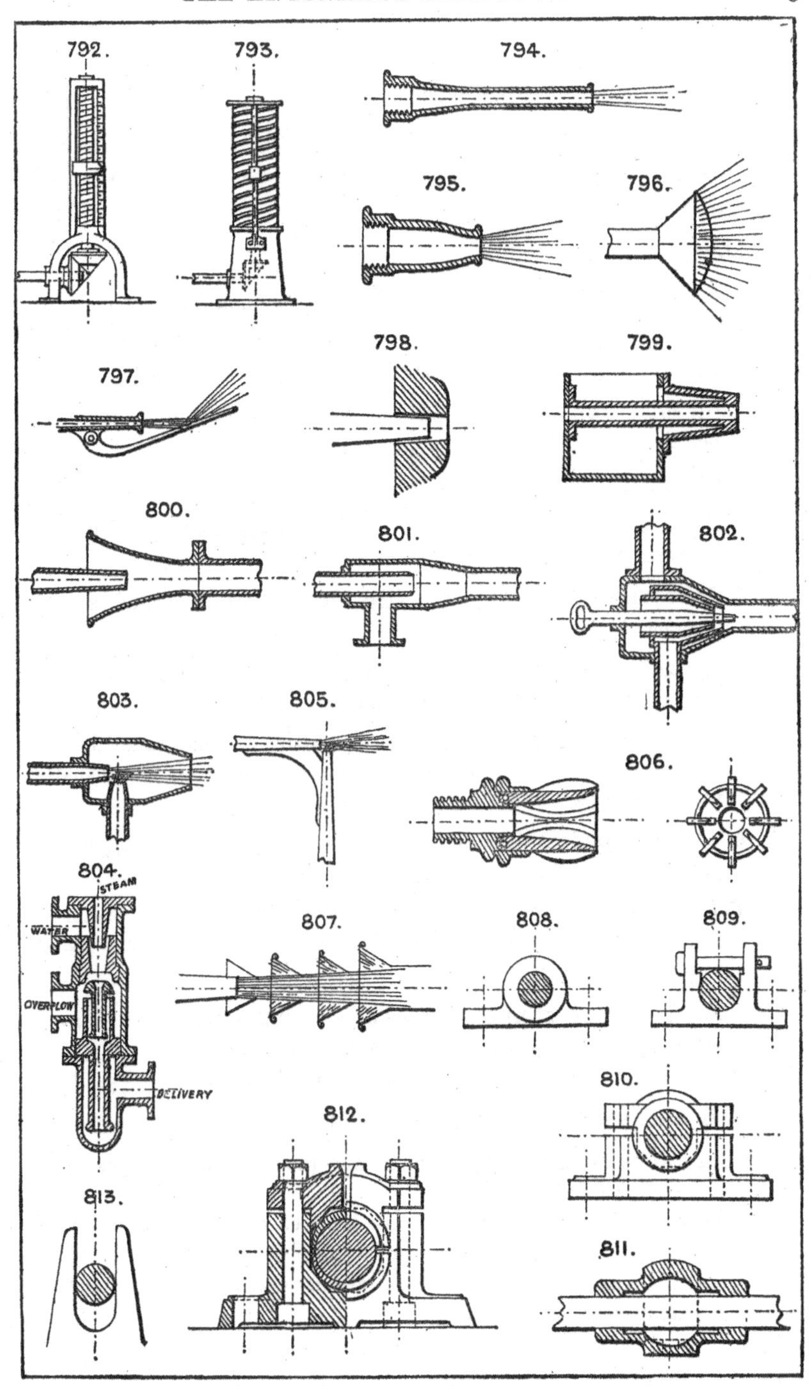
792.
793.
794.
795.
796.
797.
798.
799.
800.
801.
802.
803.
805.
806.
804.
STEAM
WATER
OVERFLOW
DELIVERY
807.
808.
809.
810.
812.
813.
811.

814 & 815. **End thrust bearings.**

816. **Sliding bearing,** with vertical or horizontal traverse.

817. **Double V bearing** to accommodate different sizes of shafts.

818. **Vertical shaft footstep.**

819. **Vertical pivot.**

820. **Horizontal pivot** and set screw; the screw should have a lock nut to prevent it being worked back by the motion of the spindle.

821. **Conical neck,** usually with steel bush.

822. **Spherical footstep,** to allow the shaft to sway out of the perpendicular.

823. **Horizontal bearing,** allowing the shaft to run out of line.

824. **Balanced bearing,** to bear the weight of a light shaft, and placed between the fixed bearings.

825. **Self-adjusting bearing** for line shafts, with ball and socket movement.

826. **Ball and socket bearing** for vertical spindle, allowing considerable variation from a right line.

827. **Horizontal thrust bearing,** with multiple flanges and double brasses, each capable of separate adjustment; used for screw shafts in steam-ships.

828. **A form of pedestal,** with the cap provided with end joggles to prevent looseness.

829. **Trunnion bearing,** for oscillating cylinders, &c. The steam is conveyed through the bearing, which has a stuffing box and gland to prevent leakage.

830 & 832. **Swinging support** for a shaft, having a sliding bevil gear or other motion upon it which has to pass the swinging support; used for lathe sliding gear, overhead travellers, &c.

831. **Ball and socket centre** for car bogies, &c.

833. **Pedestal with side adjustment** for the brasses by taper keys and screw adjustments.

Bearings running under water are generally lined with strips of lignum vitæ and require no lubricant.

So-called self-lubricator bearings are in use, lined with strips of patent composition metal.

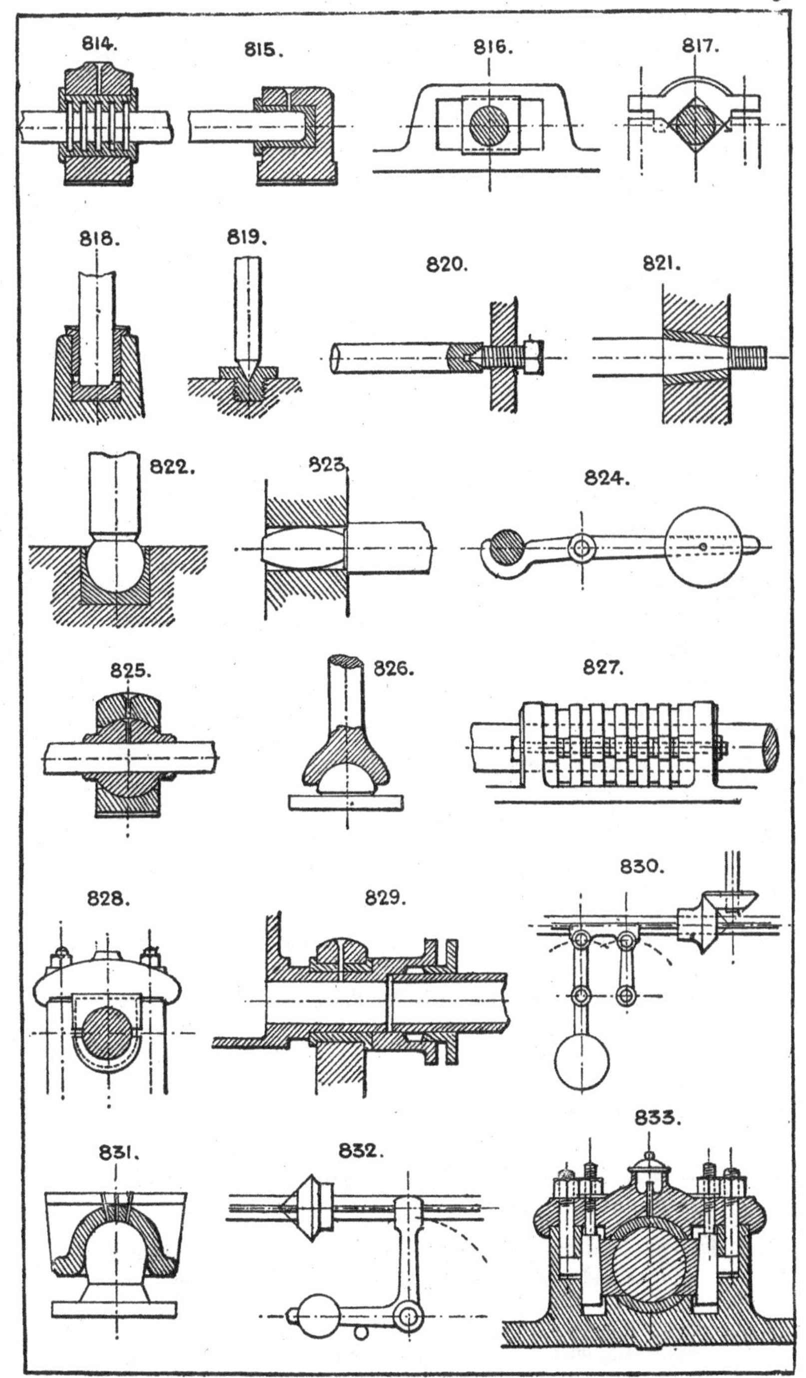
814.
815.
816.
817.
818.
819.
820.
821.
822.
823.
824.
825.
826.
827.
828.
829.
830.
831.
832.
833.

834. **Centre bearing,** with annular grip, for a heavy centre piece or car bogie.

835. **Centre bearing,** with allowance for some amount of oscillation.

Coupled bearings. No. 671.

Section 47.—PLATE WORK.

836. **Single riveted lap joint.**

837. **Double riveted lap joint.**

838. **Single riveted butt joint.**

839. **Double butt joint.**

840. **T-iron butt joint.**

841, 842, 843, & 844. **Angle or edge seams.**

845. **Transverse tubular seam.**

846, 847, & 848. **Reducing ring seams.**

849, 850, 851, & 852. **Bottom seams** round water spaces, fire-boxes, &c.

853. **Expansion hoop joint** in boiler flues, &c.

854 & 855. **Fire-box stays.**

856. **Gusset stay** for flat ends.

Flat bar, tube, and round iron stays are also much used to stay flat surfaces in boilers and tanks.

In household boilers it is usual to weld all the seams, thus avoiding L iron and other riveted work. See Nos. 89 to 96.

Flue tubes in boilers are stayed also by cross tubes inserted at intervals, such as Galloway's patent conical cross tubes.

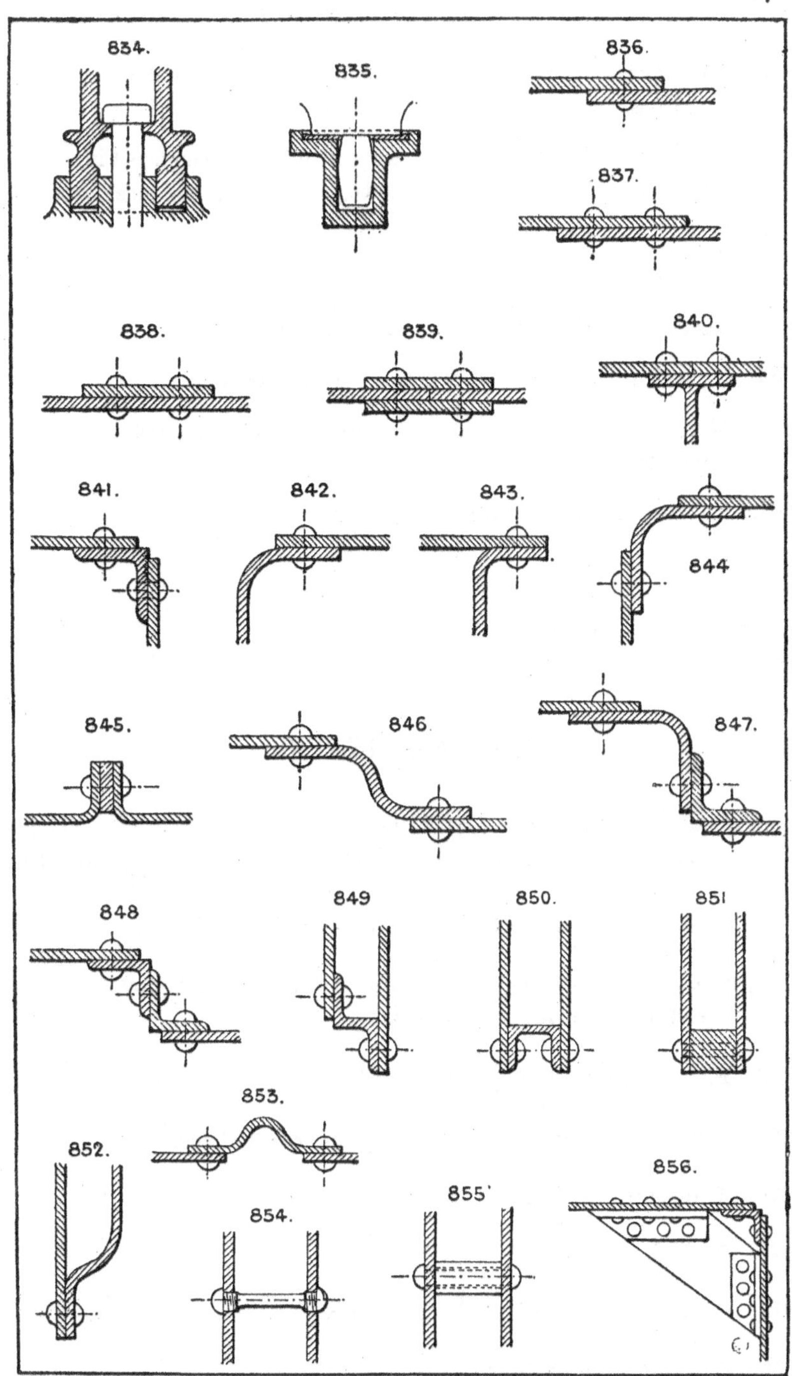
834.
835.
836.
837.
838.
839.
840.
841.
842.
843.
844
845.
846
847.
848
849
850.
851
852.
853.
854.
855
856.

857 & 858. **Cover plates** to carry tensile strains over joints in plates, L irons, &c.

859, 860, 861, 862, & 863. **Various forms of joints** employed in plate iron structures, boxes, tanks, &c., not subject to much strain. 863 is a dovetailed joint.

864. **T or L iron strut end joint.**

865. **Junction** of flat bar and diagonal T or L iron.

866. **Gusset plate joint** for diagonal ties and struts.

867. **Mode of jointing** boiler plate corners by tapering the corners of the plates.

868. **Another form of angle joint.**

Section 48.—LEVERS.

Levers are of three orders (see Section 53). The fulcrum or rocking centre may be at either end or at some intermediate point. In practice the fulcra are usually shafts or pins (see Sections 76 and 77), and the following are the typical forms in use.

869, 870, 871, & 872. **Elevation and plans of plain levers,** with end bosses for rod attachments.

873. **Plan of plain lever,** with forked end.

874 & 875. **Bell crank levers,** with plain or forked ends.

876. **T or double cranked lever.**

877. **Forked end,** off-set.

878. **Fish bellied lever** of the 2nd order.

879. **Balance weight lever.**

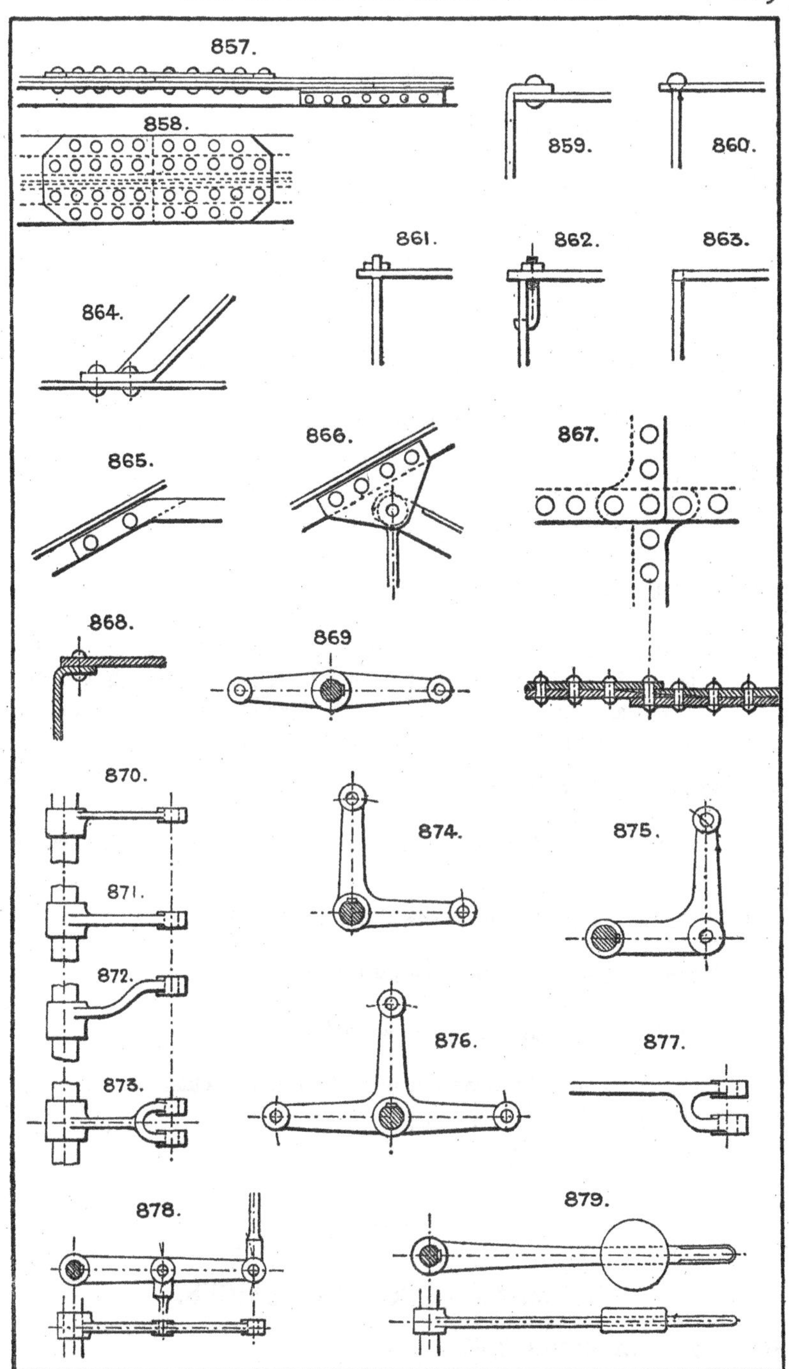
857.
858.
859.
860.
861.
862.
863.
864.
865.
866.
867.
868.
869
870.
871.
872.
873.
874.
875.
876.
877.
878.
879.

880. **Hand lever,** with round handle

881. **Hand lever,** with flat handle.

882. **Another form of round handle** sometimes used.

883. **Crank handle.**

884. **Starting lever,** with spring catch.

885. **Another pattern of ditto.**

886. **Similar lever,** with side or crank handles.

887. **Foot lever.**

888. **Foot treadle frame.**

889. **Wrist plate or T lever.**

890. **Hand lever,** adjustable as to length by means of a slot and locking bolt. For this purpose a plain round rod passed through a central socket and fixed at any radial length by a set screw, is often used; or the hand rod may be cranked as No. 1784.

891. **Double hand lever.**

892. **Lever,** formed of two wrought iron or steel plates and distance pieces.

893 & 894. **Rocking levers,** with sliding swivel joints.

895. **Forked lever,** to span a central bearing.

896. **Hand lever,** simple pattern; wrench or spanner.

897. **Headed lever,** for valve rod and other movements. See Nos. 149 to 152.

See also Section 97.

Section 49.—LOCKING DEVICES.

898. **Common sliding bolt.**

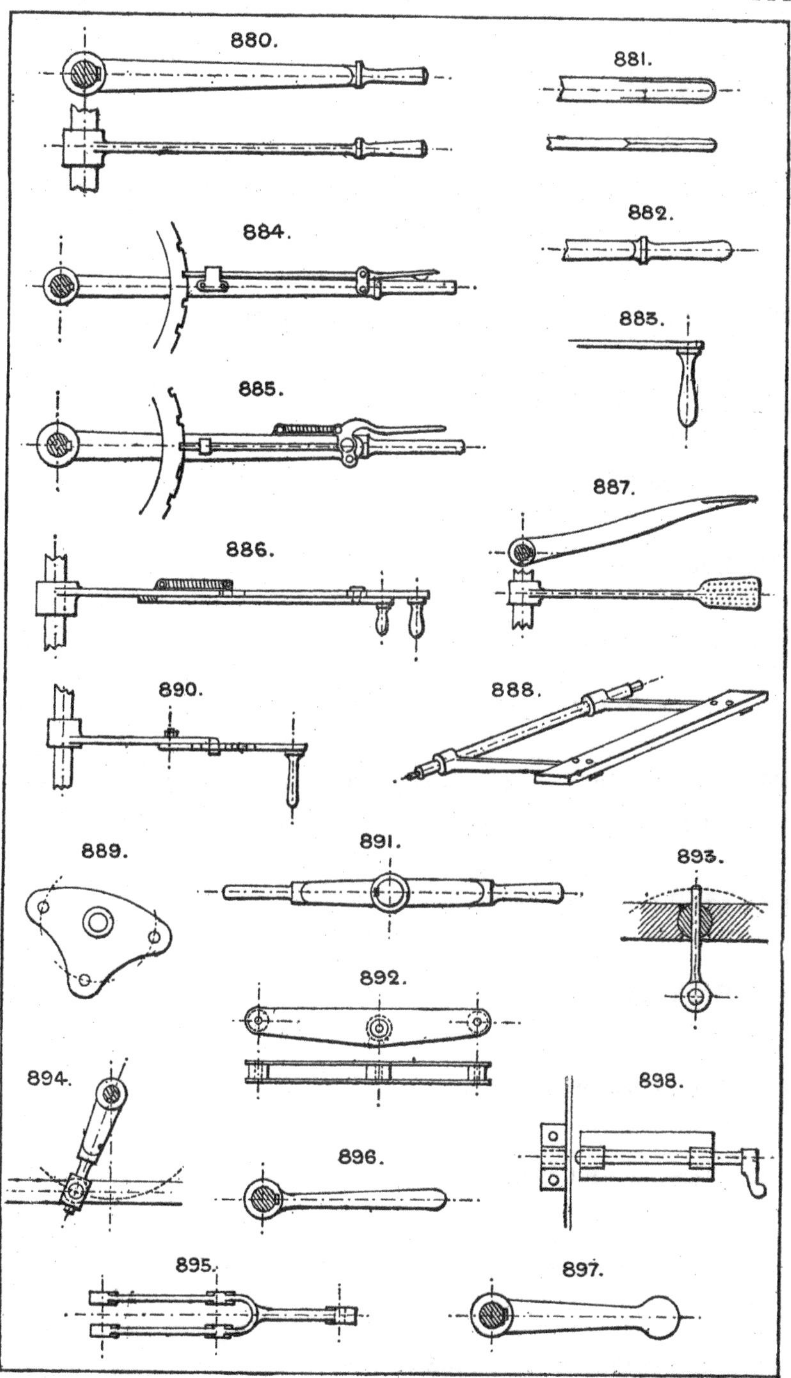
880.
881.
882.
883.
884.
885.
886.
887.
888.
889.
890.
891.
892.
893.
894.
895.
896.
897.
898.

899. **Common latch.**

900. **Cam locking bolt**; locks the bolt when either in or out, so that it can only be moved by the cam spindle.

901. **Crank movement locking bolt**, similar to the last.

902. **Bolt of common lock.**

903. **Disc and pin.**

904. **Side pawl.**

905. **Locking pawl.**

For Pawl and Ratchet Gear see Section 62.

906. **Spring catch**, with round end which slips past the socket if sufficient force is applied, used for swing doors, &c.

907. **Another form**, bevilled on one side.

908. **Hook latch.**

909 & 910. **Hasp and staple.**

911. **Crossbar and hooks.**

912. **Hand set screw.**

913. **Drop catch** for turntable, &c.

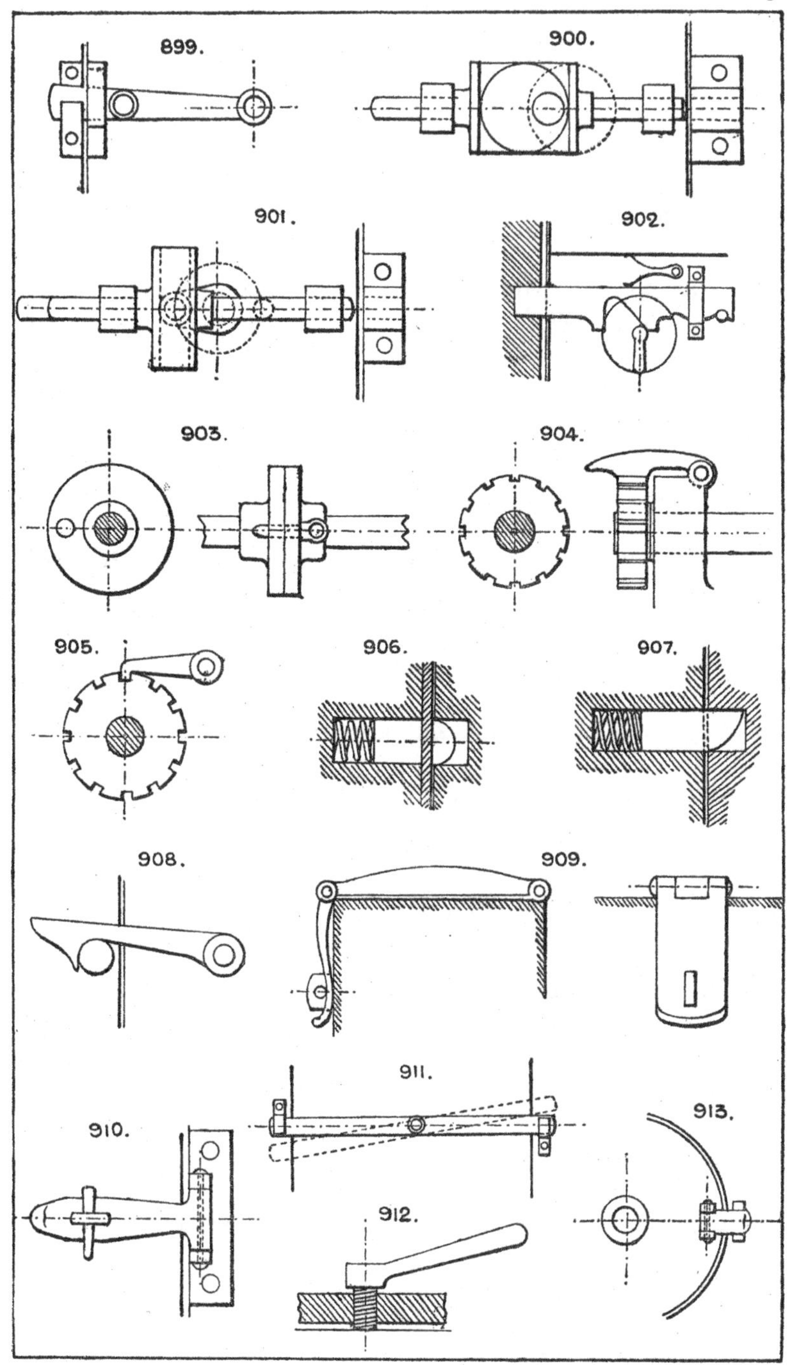
899.
900.
901.
902.
903.
904.
905.
906.
907.
908.
909.
910.
911.
912.
913.

914. **Turning or twisting bolt.**

915. **Rope or rod stopper,** with cam lever grip. See No. 47.

916. **Chain stop.**

917 & 918. **Spindle grips,** to lock a sliding or revolving spindle in a bearing or bush.

919. **Clamp and screw.**

920. **Sliding shaft locking pin;** used for lathe headstock back-gear shafts, &c.

921. **Lever locking hook;** the lever is hinged so that it can be slipped over the hook.

922. **Bow catch** for ladles, skips, &c.

923. **Segment-slot and bolt fixing** for swivelling base.

924. **Pin lock** for turntable or disc.

925. **T catch.**

926. **Roller and incline slot** for locking a rod or rope.

927. **Revolving bush lock for catch rod;** the catch rod can only slip through the bush when the latter is in one position (see plan view).

928. **Wire fencing notches** in L or ⊔ iron.

929. **Trap door automatic catch.**

930. **Screw and bridle suspension,** for blast pipes, &c.

931. **Drop loop fastening** for a door.

932. **Spring stud lock.**

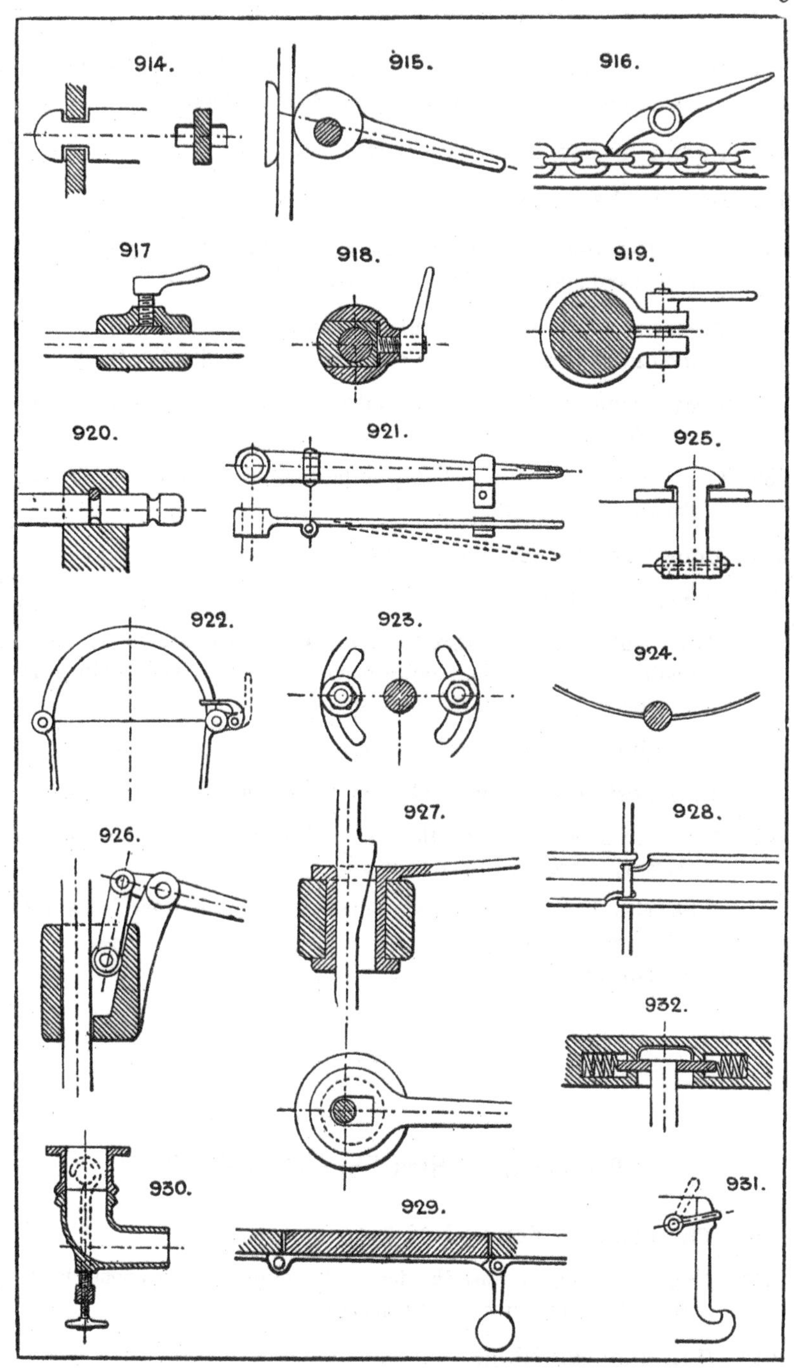
914.
915.
916.
917
918.
919.
920.
921.
925.
922.
923.
924.
926.
927.
928.
932.
930.
929.
931.

933. **Disc and radial slot**; the rod can be slipped out sideways when the disc is turned so as to bring the slots together.

934. **Radial hinged lever and crown ratchet.**

935. **Locking bar** to fix a lever in any position.

936. **Pawl for locking sliding shaft,** used for winches, &c., having double and single purchase gear or shifting clutches.

937. **Fastening eye bolt** for a hinged cover; the bolt is also hinged, and can be turned down out of the way. See also No. 1930.

938. **Crank arm device,** to lock a valve or lever in two positions. See also No. 16.

939. **Gun, breech-loading,** sliding cylindrical block locked by turning the arm into a notch.

940. **Door fastening staple and cotter.**

941. **Common cotter.**

942. **Half nut locking and unlocking device,** used for lathe leading screws; the half nuts are moved simultaneously in opposite directions by cams on the lever spindle.

943. **Swinging catch** to secure end of a drop bar.

944. **Tool post,** to swivel and lock in any position. See also No. 493.

945. **Locking screw,** to lock the hand wheel and spur pinion to the shaft when required to be driven by it.

See also Lock Nuts. The common varieties of lever locks with stepped key-wards.

See No. 1723.

Section 50.—HINGES AND JOINTS.

946. **The common double-leaf hinge.**

947. **Rising hinge,** to cause the door to lift slightly as it opens, it will then close of itself without a spring.

948. **Cup and ball hinge.**

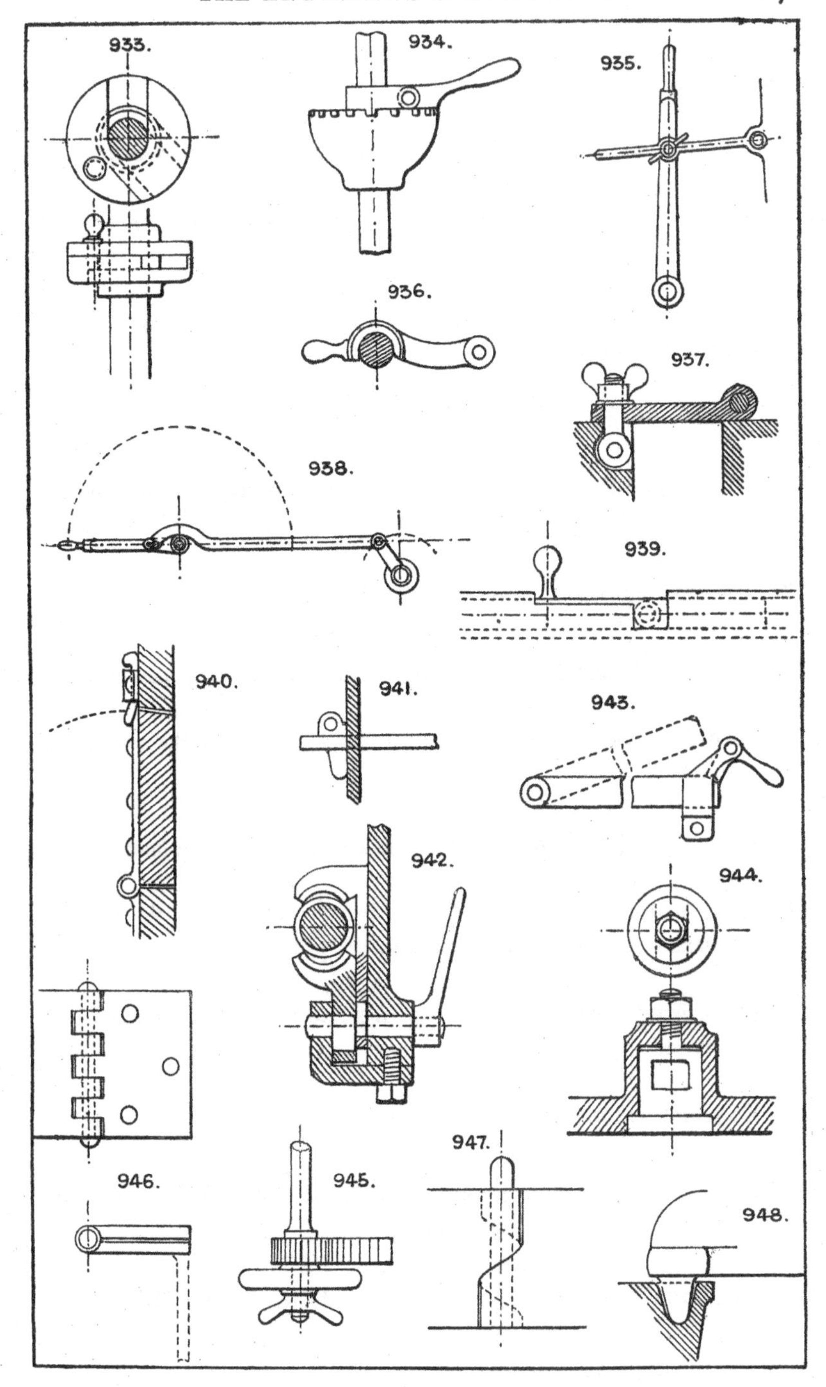
933.
934.
935.
936.
937.
938.
939.
940.
941.
943.
942.
944.
946.
945.
947.
948.

949 & 950. **Pintle hinges.**

951 & 952. **Parchment or leather hinging** for wood movements.

953. **Dovetail joint,** used on iron bedsteads, &c.; the circular dovetail is slightly tapered and fitted tight.

954. **Hinge for a door,** required to lay flat against the wall at either side when open.

955. **Hinge pin** for rocking levers on a knife edge.

956. **Door spring hinge,** to return the door always to its central position; the cams press against a roller attached to the springs.

957. **Another method,** with tension springs.

958. **Rocking bearing or knife edge,** used for weighing machines, &c.

959. **Knuckle joint,** halved together; the bolt secures the two parts together.

960. **Door spring hinge** with open springs and toggle movement.

961. **Gate hinges,** with double pintle at bottom to cause the gate to return to the central position without springs.

962. **Link hinge,** for a grid or trap door, to allow it to lie flat when opened.

963. **Bayonet joint.** A common device.

964. **Double scarfed and joggled joint,** for pump, rods, &c., with ferrules and keys to tighten up.

965. **Universal joint.** See Dr. Hooke's Joint, Nos. 33 and 34.

966. **Knuckle jointed levers.**

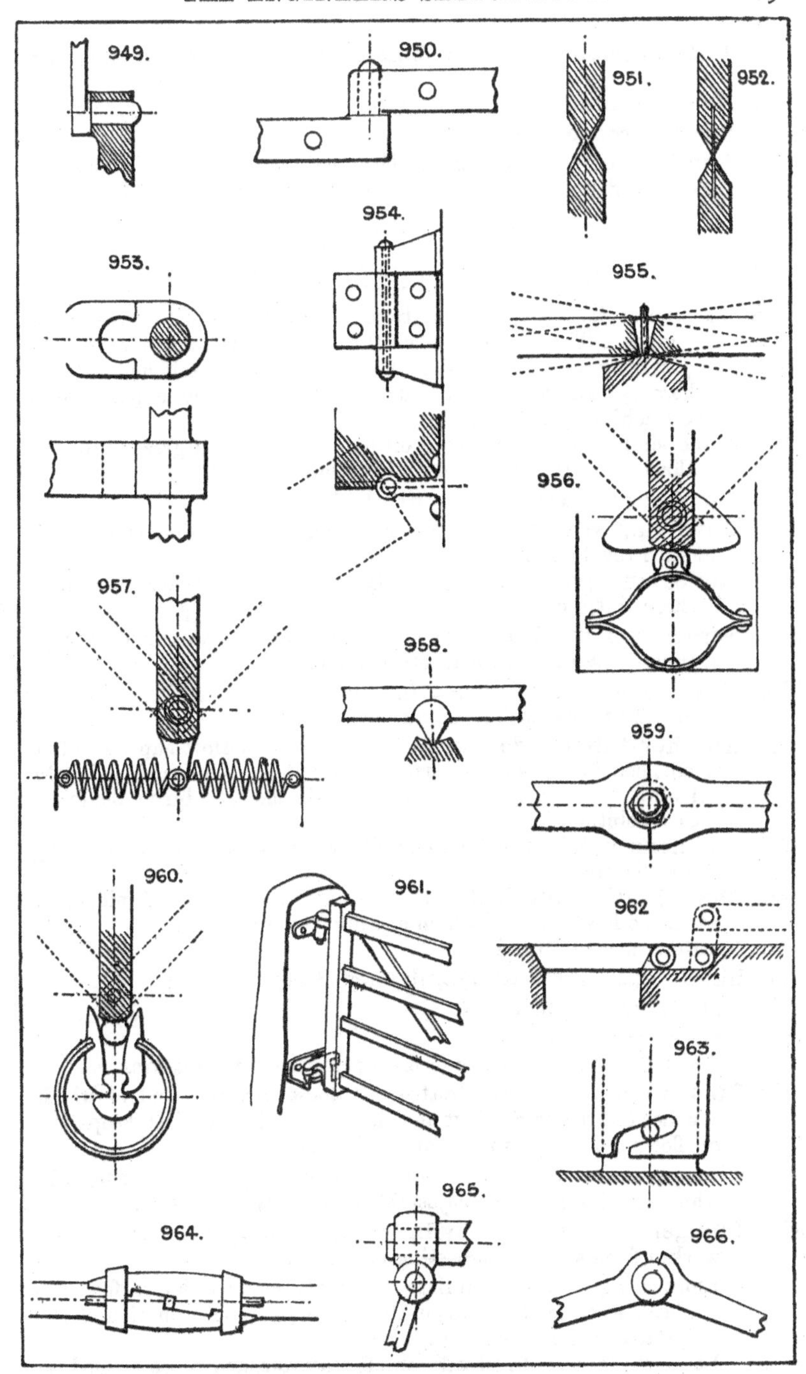
949.
950.
951.
952.
953.
954.
955.
956.
957.
958.
959.
960.
961.
962
963.
964.
965.
966.

967. **The common male and female or nipple and socket rod joint.**
968. **Multiple hinges,** with one centre bolt, for long or heavy doors.
969. **Scarfed rod or bar joint.**
970. **Another form of hinge,** to effect the same object as No. 954.
See also Swivel Joints, Nos. 893 and 894. Sections 49, 4, and 48.

Section 51.—LUBRICATORS.

I do not propose to attempt to illustrate the vast tribe of "greasers" of all kinds. They would easily fill a moderate volume, but scarcely repay the reader for perusal. I shall content myself here as elsewhere by indicating the types of most interest and importance to the machine draughtsman.

Besides the simple cup or enlarged oil hole, oil box, and grease cup, the following are the most commonly employed:—

971. **Oil pan** for gearing, worms, wheels, &c.
972. **Revolving wire lubricator**; carries a drop of oil on to the shaft at each revolution.
973. **Roller and pan lubricator.** Can be employed also for gum, paste, paint, &c.
974. **Screw ram lubricator,** to force lubricant into a cylinder or pipe against pressure, with non-return valve.
975. **Telescopic tube lubricating device,** for reciprocating or revolving joints, such as crank pins.
976. **Another tubular device** for crank pins; a hollow cup on end of a tube stands opposite the centre of the shaft, and can be fed with oil while revolving, the oil running down the tube during the lower half revolution.
977. **Stauffer's lubricator** for thick oil, which is forced in by screwing down the cap.
978. **Shaft bearing lubricator** by the capillary action of pieces of cane, the lower ends of which dip into the oil cistern.
979. **Endless string lubricator.**
980. **Single cock lubricator,** with screwed cap for filling.
981. **Double cock lubricator.**
982. **Hollow plug cock.**
The last three are used to feed oil against steam pressure.
983. **Lieuvain's needle lubricator.** A loose wire (one end of which touches the revolving shaft and the other is in the oil); keeps the oil flowing as long as the shaft is running.
984. **The pressure of steam enters the cup above the oil,** which is fed through an adjustable small valve at bottom.
985. **Plunger or ram and cylinder lubricator,** with ratchet feed worked from some reciprocating part of engine.

Lubricating or inking rollers to evenly cover a flat surface are placed at an angle of about 10° to the direction of motion of the surface to be lubricated or inked.

Large engines are fitted with an oil reservoir, and pipes are led to all joints, bearings, &c., with small cocks for regulation.

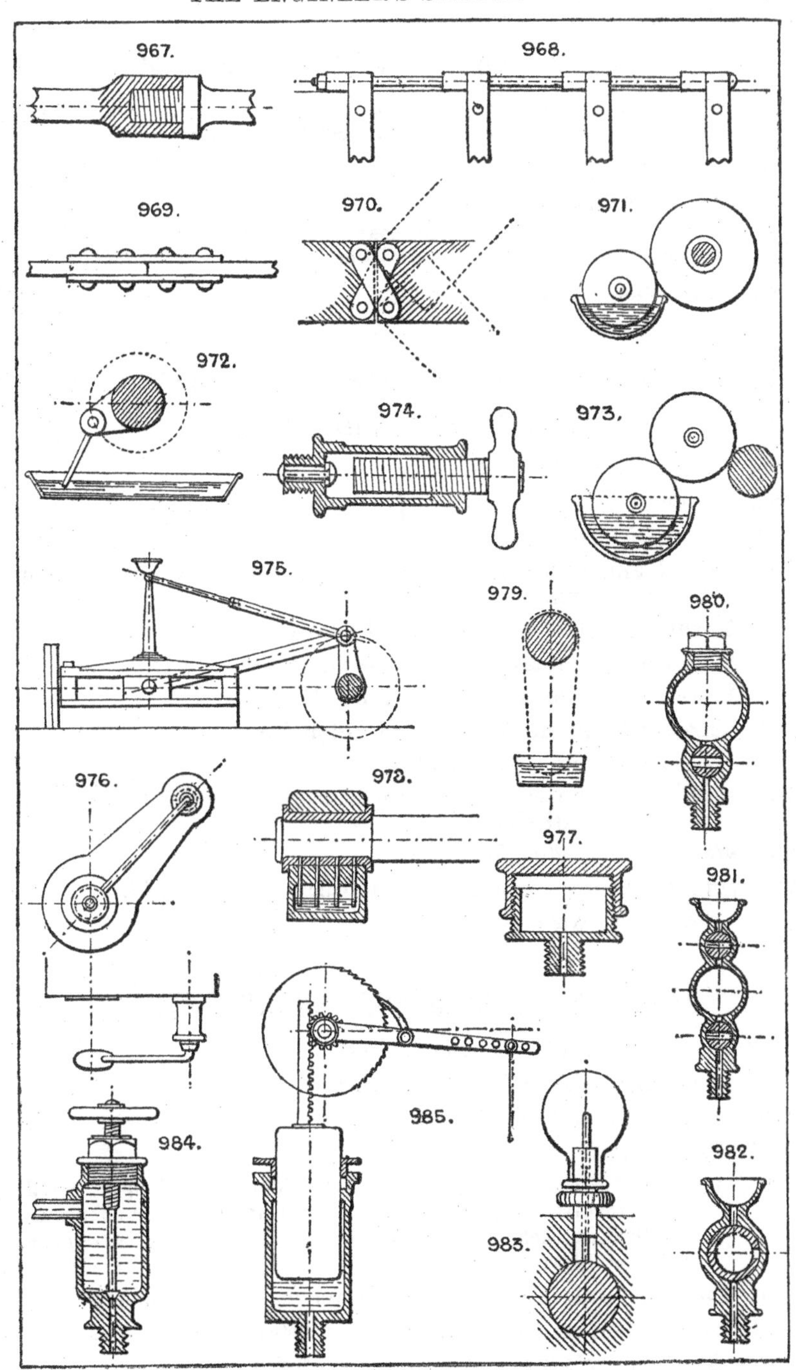
967.
968.
969.
970.
971.
972.
974.
973.
975.
979.
980.
976.
978.
977.
981.
984.
985.
983.
982.

Section 52.—LEVELLING AND PLUMBING.

The common spirit level and plumbline are ordinarily employed, as also the surveyor's telescope and spirit level or "Dumpy."

986. **Gravitation level.**

987. **Plumbline and square.**

988. **Water tube level**; the tube may be carried a long distance and round corners, &c., in any direction below the water level.

989. **Spirit level plumbing square.**

Section 53.—MECHANICAL POWERS. APPLIANCES TO VARY POWER AND SPEED.

990. **Lever of the 1st order**
991. **Lever of the 2nd order**
992. **Lever of the 3rd order**

} See Applications, Section 48.

993. **Wheel and axle**; power gained in proportion to the diameters of the two sheaves.

994. **Return block**; power multiplied 2 to 1.

995. **Two double sheave blocks**; power multiplied 4 to 1.

996. **Four single blocks**; power same as No. 995.

997. **Three return blocks**; power multiplied 8 to 1.
See Applications, Section 42.

The inclined plane is simply a modification of the force of gravity, which acts vertically.

The wedge. See Section 37.

The screw is simply a circular inclined plane. See Section 78.
See also Gearing, Section 40.

Section 54.—MIXING AND INCORPORATING.

998. **Kneading mill,** with spiral vanes.

999. **Pug mill,** with radial spiral paddles.

1000. **Pug mill,** with spiral paddles.

1001. **Pan mixer.**

1002. **Egg beater or mixing machine.** Two sets of open radial frames revolve in opposite directions, the frames being shaped to pass through each other, and are driven by mitre gear driving a shaft and sleeve.

1003. **Diagonal mixing barrel,** with revolving and fixed vanes.

1004. **Conical mixing barrel** of similar construction.

1005. **Diagonal mixing pan,** used for confectionery, &c.

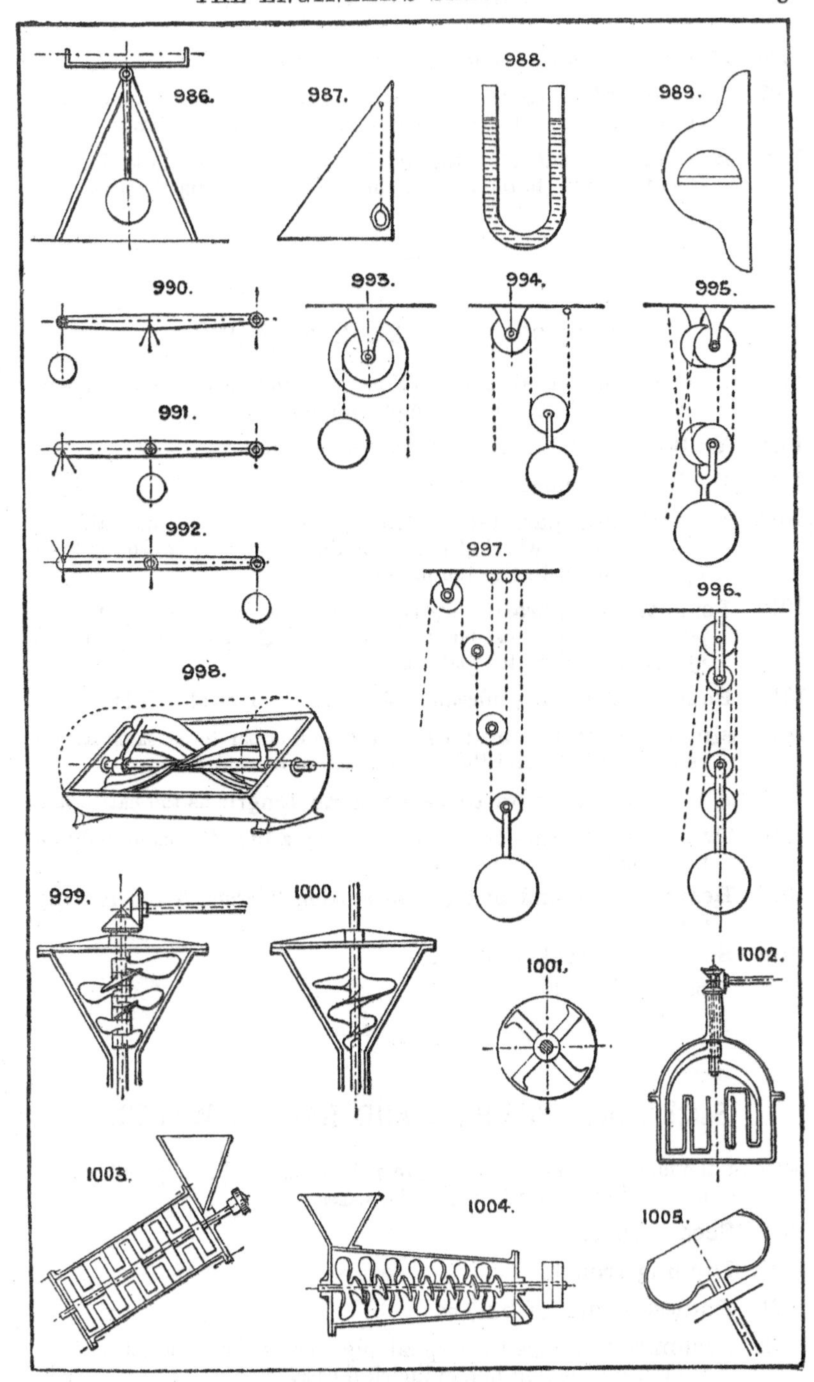
986.
987.
988.
989.
990.
991.
992.
993.
994.
995.
996.
997.
998.
999.
1000.
1001.
1002.
1003.
1004.
1005.

1006. **Mixer,** with two pairs of arms running in opposite directions.

1007. **A modification of the last;** the centres of the arms being above one another so that the arms pass each other in revolving.

1008. **Horizontal table mixing machine.** The stuff works its way from the centre to the edge of the table by centrifugal force.

Section 55.—PARALLEL MOTIONS.

Purpose: to maintain rectilinear motion of a rod or equivalent detail coupled to a lever without employing guide bars.

1009. **Watt's parallel motion** for a beam engine.

1010. **Rack and segment motion.**

1011. **Epicycloidal parallel motion.** The pinion is one-half the diameter of the wheel at the pitch lines, and the gudgeon is fixed upon the pitch line of the pinion.

1012. **Peaucellier's parallel motion.** A is a fixed centre; B (for *parallel* motion) must be one-half way to C; power is applied to C; D parallel centre gudgeon.

1013. **Beam,** with rocking fulcrum. A A are equal, as also B B.

1014. **Single radius bar and link;** the radius bar to be same length as the half beam and the link hinged on its centre.

1015. **All the radius bars to be of same length** as the half beam.

1016. **Two equal radius bars** connected by a link, the main gudgeon in its centre.

1017. **Beam of the 2nd order** with rocking fulcrum, A and A being equal.

1018. **Sector and rack motion.**

See also No. 714.

Section 56.—PUMPING AND RAISING WATER.

Some of the most primitive methods are still in use, and may possibly still be found of service in particular cases.

1019. **Scoop wheel.**

1020. **Dipping trough.**

1021. **Endless chain of buckets.**

1022. **Archimedean screw;** a spiral pipe serves the same purpose as the worm revolving in a cylindrical case.

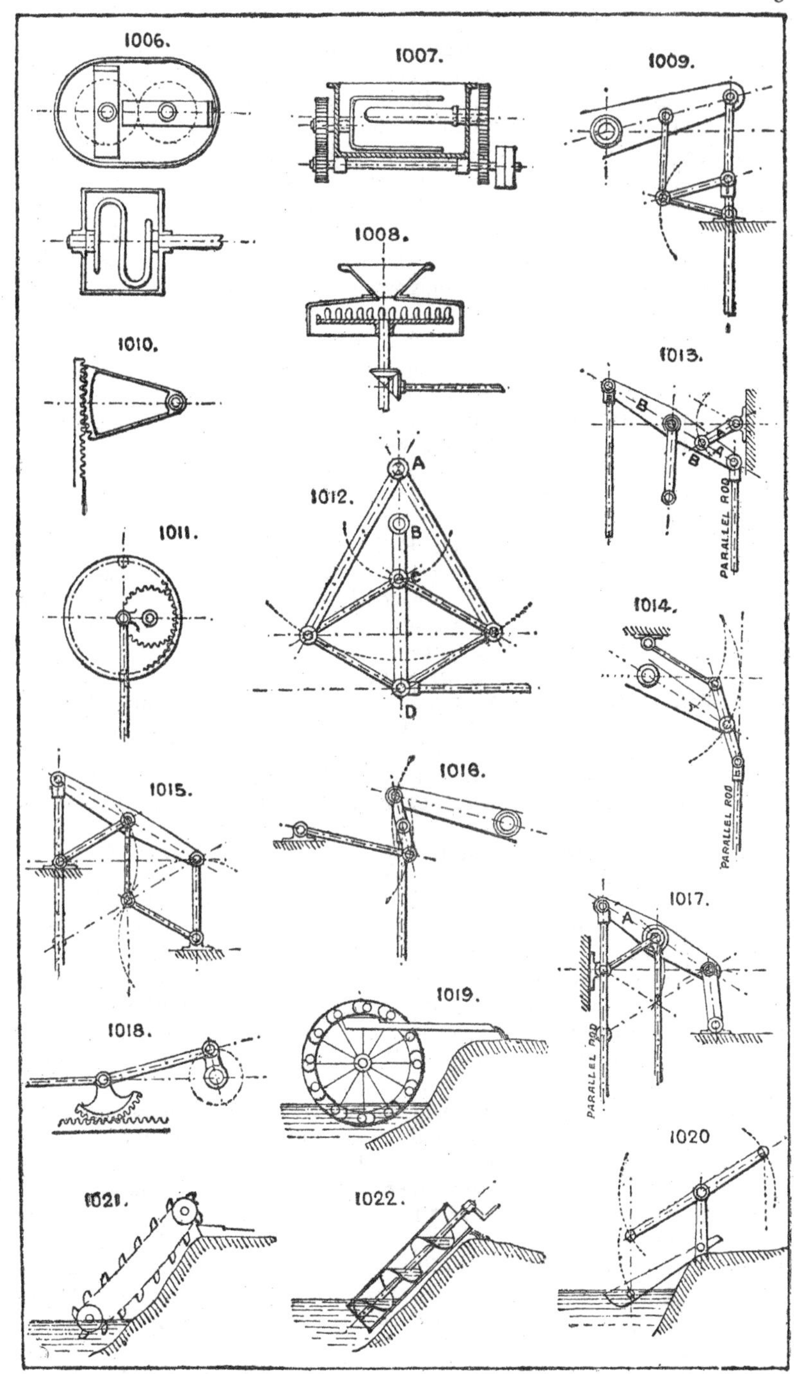
1006.
1007.
1009.
1008.
1010.
1013.
B
A
A
B
PARALLEL ROD
A
1012.
B
C
D
1011.
1014.
PARALLEL ROD
1015.
1016.
1017.
A
PARALLEL ROD
1018.
1019.
1020
1021.
1022.

1023. **Chain pump,** frequently used still. The lower length of pipe should be bored to fit the buckets on chain; the rest of the pipe may be a little larger in diameter and not bored.

1024. **Lifting wheel** for raising water.

The following four examples are machines for raising water to any height by employing a fall of water of comparative low pressure:—

1025. **Hydraulic ram.** A stream of water runs down the incline pipe and flows away at the ball valve; when its speed reaches a certain point it suddenly closes the ball valve, and the shock opens the delivery valve, water flows into the air vessel till the power of the stream is checked, when the delivery valve closes, the ball drops, and the action is repeated.

1026. **The Robinet.** Direct-action water pressure self-acting pump; performs the same work as the hydraulic ram, that is by using a low fall and large quantity of water it raises a smaller quantity to a greater height, the low-pressure water acting on the large double-acting piston. The valve is reversed by the motion of the engine.

1027. **Hydraulic pumping engine.** A modification of the Robinet A is the driving cylinder, B the pump. The main slide valve is worked by two pistons, and the pressure water distributed by an auxiliary four-way cock or small slide valve, connected to a stop rod from the main crosshead. See also No. 1741. See Sec. 93.

1028. **Water wheel and pump.**

1029. **Single-acting bucket or suction pump.**

1030. **Single-acting ram force pump.** Sometimes an open top cylinder and piston are used instead of a ram, as No. 1029.

1031. **Double-acting ram and piston pump.** Forces on both strokes; sucks only on the up stroke.

1032. **Double-acting plunger or ram pump,** externally packed; a favourite arrangement.

1033. **Double-acting piston pump,** four valves. This is of course a type of a very great variety of pumps.

1034. **Double-action piston pump,** without valves. The piston has an oscillating or radial motion (see plan), as well as an up and down motion, so that the two ports are alternately open to the upper and under side of piston by the small passages A A. The required motion can be obtained from No. 406, crank motion.

1035. **Rope pump.** A simple endless soft or porous rope absorbs water at its lower part (immersed), which water is pressed out of it between the rollers at top.

1036. **Apparatus to supply air to air vessels.** The main pump at every stroke draws a small quantity of water from the small air vessel A, and on the return stroke forces an equal quantity of air from the smaller to the larger vessel B. C is a double air valve.

1037 **Combined bucket and plunger pump** draws on the up stroke only, but delivers on both strokes.

1038. **Air pump,** with foot and head valves.

1039. **Hand or power pump.** Can be thrown into action from the crank shaft by fastening the set screw, or may be worked independently by the hand lever.

1040. **"Worthington" pattern of plunger pump,** double acting.

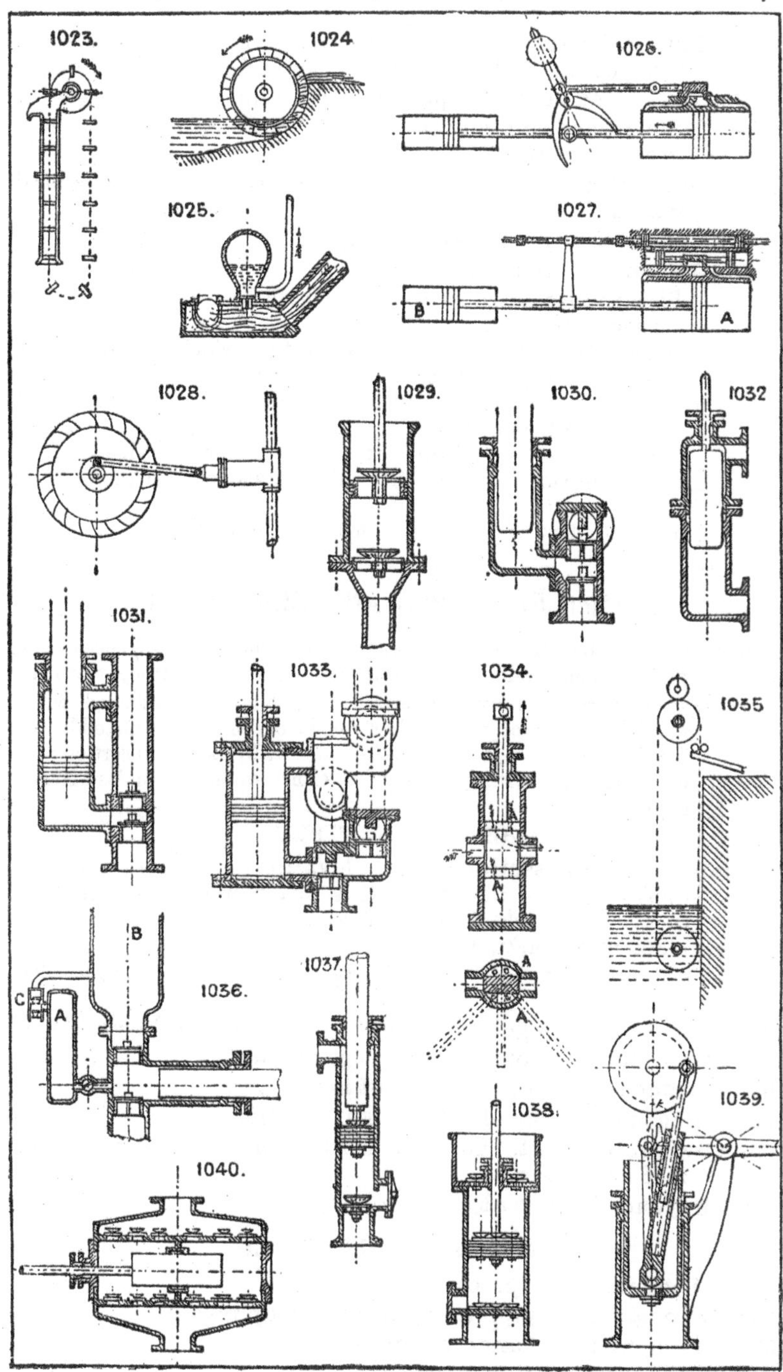
1023.
1024
1025.
1026.
1027.
B
A
1028.
1029.
1030.
1032
1031.
1033.
1034.
A
1035
B
C
A
1036.
1037.
1038.
1039.
1040.

1041. **Double barrel pump**, with bucket pistons. The water passes both pistons, which are fitted with valves opening opposite ways.

1042. **Oscillating sector** or quadrant pump, with one vane or piston.

1043. **Double quadrant pump.** The two vanes are worked by links from a single crank.

1044. **Oscillating pump**, with two radial vanes keyed to a central rocking shaft.

1045. **Hollow plug**; oil or water feeder.

1046. **Double ram pump.**

Rising mains in mines and wells have been used as the main pump rods in some instances.

In pumping up to a tank or reservoir from which a down service pipe is taken, the pump can be arranged to deliver into this pipe at its nearest convenient point, instead of having a second pipe from the pump to the tank.

Schmid's trunk cyl. hydraulic pumping engine utilises a low-pressure supply to force from the annular side of the piston a high-presure service.

See also rotary engines and pumps, Section 75.

Section 57.—PIPES AND CONVEYORS.

Plain tubing may be either of iron, brass, zinc, lead, tin plate, sheet iron, papier mâché, indiarubber, guttapercha, leather, cotton, or canvas.

Flexible sorts of the last five materials named are strengthened when required by spiral wire, either inside or outside or imbedded in the material; also, in the case of rubber, by canvas insertion; or by being payed with yarn or wire, either wound or plaited round the exterior.

CAST IRON PIPES.

1048 to 1053. **Show sections and elevations** of forms of flanges employed. 1053 **has a small V space**, in which is inserted a ring of guttapercha cord or soft lead. Used for heavy pressures.

1054. **Socket and spigot pipes.**

See also the ordinary earthenware socket drain pipes, flue pipes, &c.

1055. **Socket and spigot pipes**, with tapered, bored, and turned joint.

1056. **Cup and ball joint** for uneven ground, &c.

1057. **Wrought iron pipes**, with cast iron flanges.

1058. **Diagonal universal joint.** See No. 1078.

1059. **Swivelling joint.** Coupling quickly opened or closed, faced with rubber or leather.

1060. **Bayonet joint**, for hydrants, &c.

WROUGHT IRON PIPES.

1061. **Sheet iron flue pipes.**

1062. **Wrought iron pipe and screwed couplings.**

1063. **Wrought iron flange coupling.**

1064. **Reducing socket** or coupling.

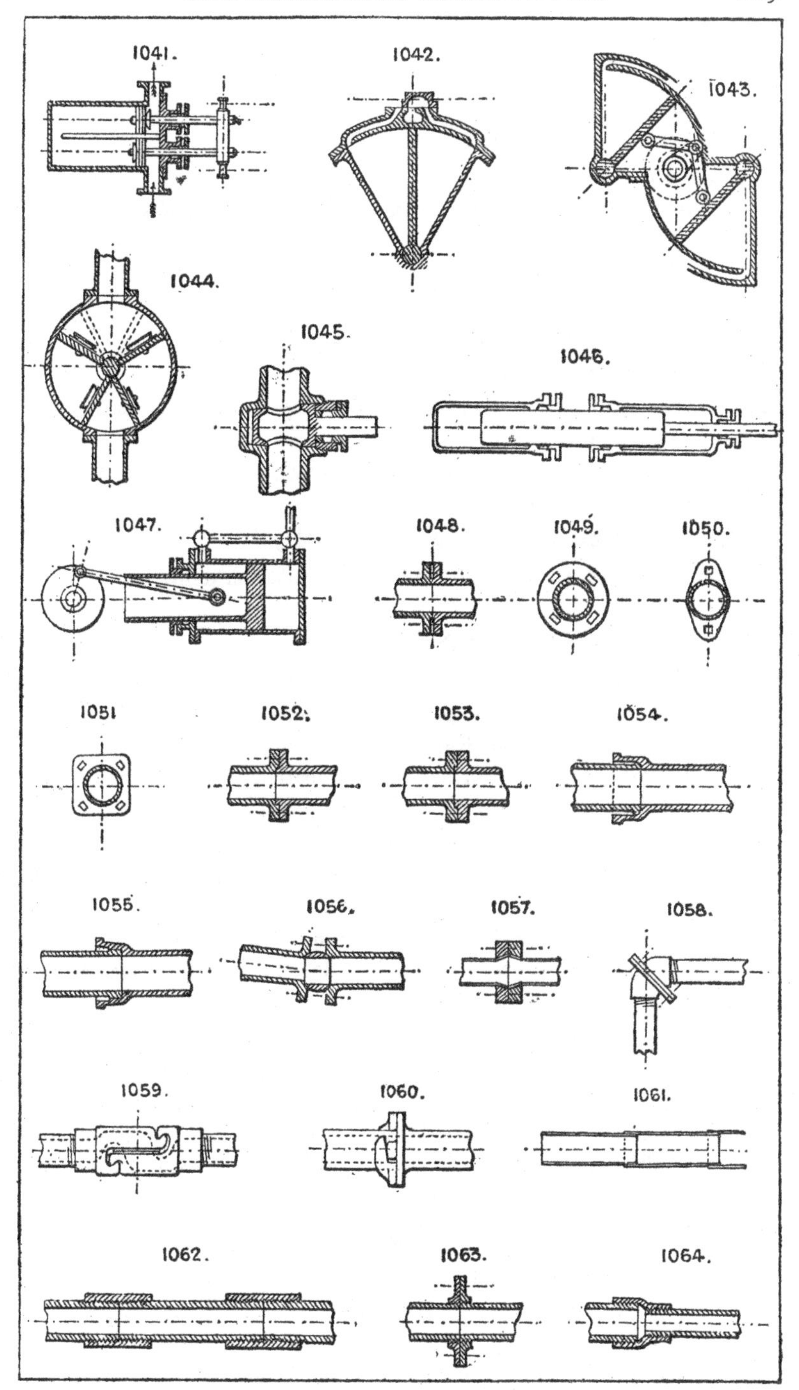
1041.
1042.
1043.
1044.
1045.
1046.
1047.
1048.
1049.
1050.
1051
1052.
1053.
1054.
1055.
1056.
1057.
1058.
1059.
1060.
1061.
1062.
1063.
1064.

1065 & 1066. **Elbows.**

1067. **Bend.**

1068. **Internal coupling** for handrailing, &c.

1069. **Patent lap-folded pipe.**

1070. **Long screw coupling and back nut,** for making the last joint in a series of pipes when the last piece cannot be screwed into both joints.

1071 to 1073. **Screwed unions.**

1074 & 1075. **Unions** with right and left-hand threads.

1076. **Expansion joint** plain.

1077. **Expansion joint** with gland and safety bolt to prevent the joint blowing out.

A bent U-shaped tube of copper is sometimes used as an expansion piece in a line of hot piping.

1078. **Royle's patent diagonal universal joint.**

1079. **Expanding pipes** with stuffing boxes at each joint.

CONVEYORS.

1080 & 1081. **Wood troughs** sometimes lined with metal.

For conveying materials other than liquids, such as sand, coal, grain, &c., the following contrivances are used :—

Endless bands of canvas, rubber, leather, &c., sometimes with flanges like 1082.

Sloping wooden tubes or shoots.

1082. **Sectional conveyor,** endless, carried round pulleys like No. 1083.

1083. **Creeper,** an endless chain of boards or buckets sliding along a wood trough. See Ewart's patent detachable drive chain, &c.

1084. **Worm and trough,** similar in principle to an archimedean screw. See No. 1022.

Elevators for vertical or sloping conveyance usually consist of an endless band of some flexible material or chain with a number of tin or metal buckets attached at regular intervals like a creeper, No. 1083, but working in an enclosed tube.

Pneumatic tubes. See Section 19.

1085. **Is an improved form of worm** having no centre shaft.

1086. **Elevator,** or band and buckets.

1087 & 1088. **Endless web and roller devices** for conveying sheets of paper, for printing or folding.

See also Raising and Lowering, Section 69.

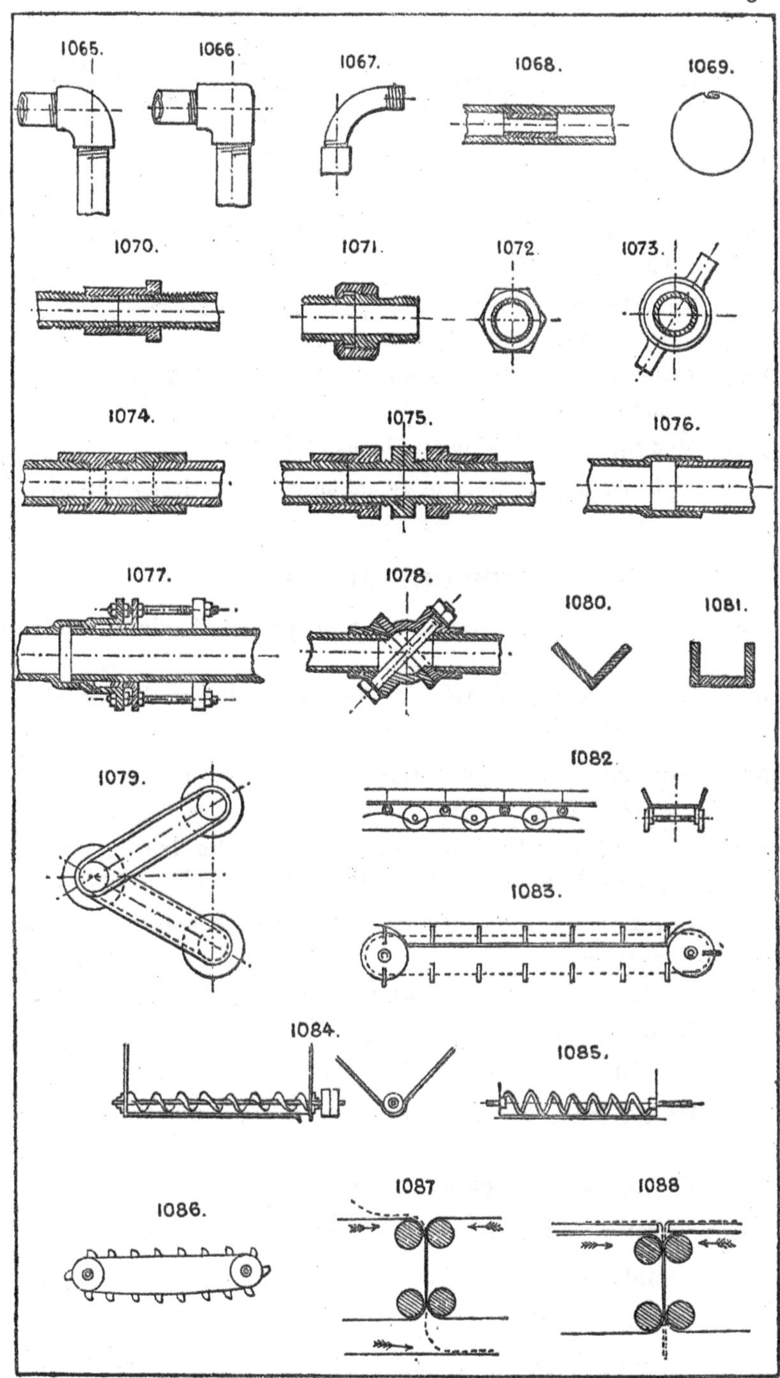
1065.
1066.
1067.
1068.
1069.
1070.
1071.
1072.
1073.
1074.
1075.
1076.
1077.
1078.
1080.
1081.
1079.
1082.
1083.
1084.
1085.
1086.
1087
1088

Section 58.—PACKINGS, JOINTS, STUFFING BOXES, &c.

PISTONS.

1089. **Piston with junk ring**; the packing is sometimes cast iron, steel, brass, or phosphor bronze rings, or even hemp or asbestos.

1090. **Small pistons** have generally two rings of steel or brass sprung into the grooves.

1091. **Double-acting hydraulic piston** for *cold* water; if single acting, one leather only is required.

1092. **Indiarubber rolling ring packing,** used on piston water meters.

1093. **Piston,** with junk ring for fibrous packing.

Numerous patents are in use for springs of various kinds applied to piston rings. See Section 80.

STUFFING BOXES, &c.

1094, 1095, 1096, & 1097. **Sections and plans of gland stuffing boxes.**

1098. **Leather packing ring or collar,** used generally for higher pressures up to 3 or 4 tons per square inch.

1099. **Stuffing box for hydraulic rams,** up to pressures of about 1000 lbs. per square inch.

1100 & 1101. **When the wear on an hydraulic leather is considerable a guard-ring should be added,** as shown here.

1102. **Stannah's patent stuffing box**; the packing is tightened by a set screw.

1103 & 1104. **Ram leathers,** with gland to facilitate renewing.

1105. **Grooved steam packing.** It is said that the steam will not readily pass a series of grooves round a piston rod.

1106. **Useful form of gland** for a screwed spindle, the thread being cut in the outer cap.

1107. **V-ring piston packing rings.**

1108. **"Bottle" gland** to cover a reciprocating rod end.

1109. **Gland** with oil space to keep the rod lubricated.

1110. **Water lute or seal.**

1111. **Grooved joint** for packing round covers, &c.

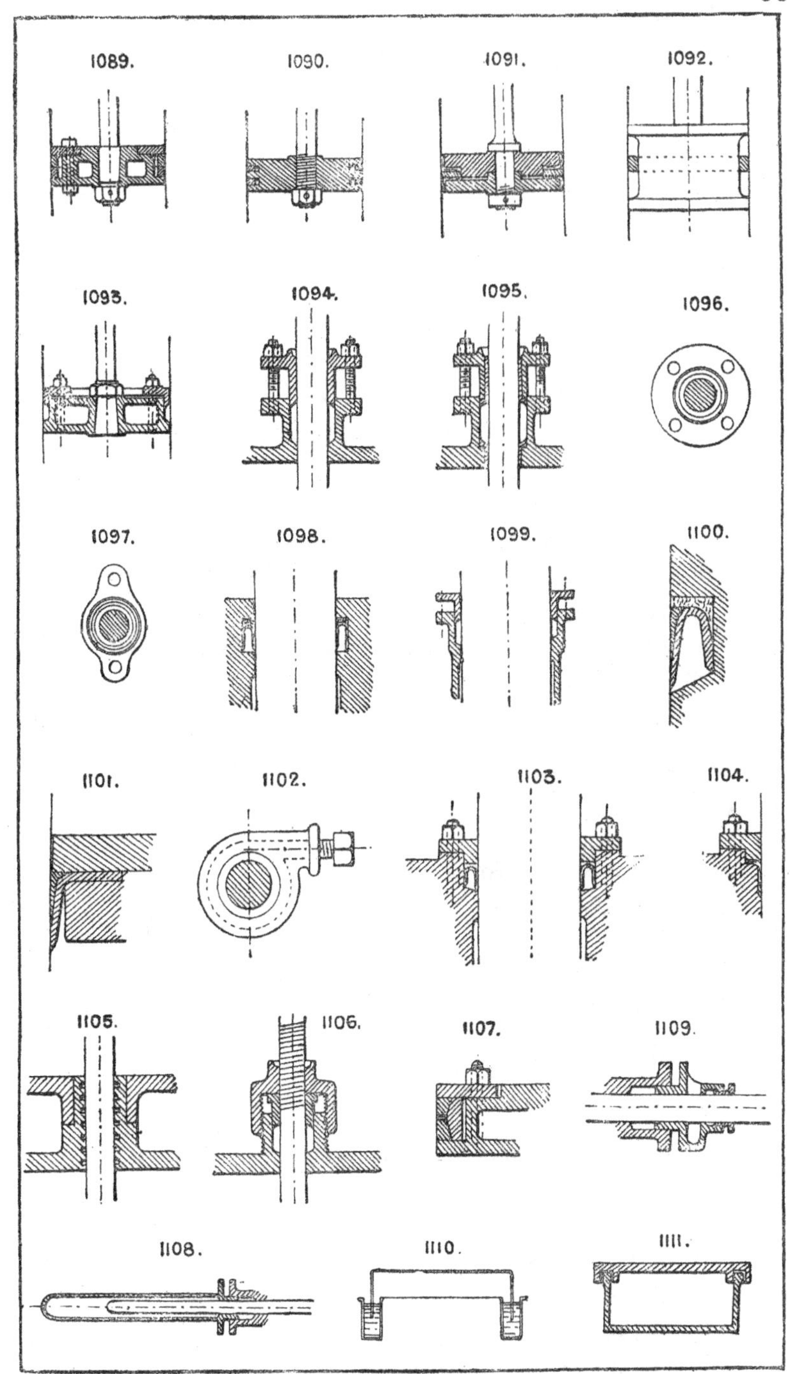
1089.
1090.
1091.
1092.
1093.
1094.
1095.
1096.
1097.
1098.
1099.
1100.
1101.
1102.
1103.
1104.
1105.
1106.
1107.
1109.
1108.
1110.
1111.

1112. **Indiarubber sheet joint** for the tubes of condensers, the rubber being pressed around the tube joints by a plate with projecting rings. See Plate, page 139.

1113. **V-ring metallic gland packing.** See Plate, page 139.

Joints of plane surfaces are usually made with red-lead for steam and water; asbestos, millboard, and sometimes rubber insertion or wire gauze for steam, water, air, &c.

Section 59.—PROPULSION.

(See also Sections 53, 60, and 12.)

On land, vehicles, &c., may be propelled by:

a. **Any engine** having contained in itself its source of power; such as a steam engine, compressed-air engine, electric motor, &c.

b. **Any fixed source of power,** the moving vehicle being connected to it by: 1, a rope or chain; 2, a tube; 3, an electric wire or other electric connection.

c. **By gravitation** down an inclined or vertical road. See Section 69.

d. **By wind power,** using sails or windmill. See Section 95.

e. **Animal power.**

On water, vessels are propelled by:

a. **Wind.**

b. **Steam** or other heat engine.

c. **Wave motion.**

d. **Natural currents,** tides.

e. **Animal power.**

In air, balloons have been propelled by:

a. **Wind.**

b. **Some kind of engine power.**

c. **Hand power.**

But the two latter sources must be at present considered almost impracticable.

MEANS EMPLOYED FOR PROPULSION.

On land:

a. **Steam or other engine and boiler** on the moving vehicle.

b. **A reservoir of compressed air** or gas, driving an engine on the vehicle.

c. **An electric battery** or accumulator, driving an engine on the vehicle.

d. **Rope railway:** the rope may be driven by any kind of engine.

e. **Endless rope** transmission. See Section 66.

f. **Inclined or vertical hoists.** See Section 69.

g. **Ice vessels,** or yachts, windmills, &c. See Section 95.

h. **Velocipedes** of all kinds, hand power lifts and hoists. See Section 69.

On water, vessels are propelled by:

a. **Sails.**

b. **Steamships,** by screw, paddle-wheel, stern wheel, water jet, and steam jet.

c. **Wave engine.**

d. **Barges and rafts** usually employ tidal motion only.

e. **Rowing boats,** &c., hand power paddle and screw boats, horse towage.

In air, balloons are propelled by:

a. **Wind,** acting on the inflated balloon, or on an umbrella-shaped or other sail; also on the under side of inclined planes of large area.

b. **Balloons** of elongated form have been propelled by an engine placed in the car, driving either a large screw propeller or wings.

c. Various attempts have been made to work flying machines (generally having some form of wings) by the power of a man's hands and feet, with very little success.

Section 60.—MOTIVE POWER.

All physical energy is assumed to be derived more or less directly from the sun, whose rays combine: 1, heat; 2, light; 3, actinic or chemical power.

Heat may be obtained:

a. By direct use of the sun's rays.

b. From any combustible material.

c. From chemical reaction.

Light does not separately develop power.

Chemical reactions are employed to develop heat, combustion, contraction, or expansion, as means of developing power.

From the foregoing elementary physical sources the following are the practical sources of our power for mechanical purposes.

Electrical power.

Magnetic power.

Tidal motion.

Falling water.

Descending weights.

Wave motion.

Wind.

Expansion of air or other gases.

Steam.

Explosives.

Fuels, hydrocarbons, &c.

These are employed in producing power by the following apparatus or motors:—

Electric motors driven from a dynamo, battery, or accumulator.

Magnetic power cannot be employed continuously as a motor, as it gives out only as much as it receives.

Tidal motion can be utilised to drive any kind of wheel, see Water Wheels, Section 90. It can also be stored in a reservoir, driving a water engine as it flows in and out on the flood and ebb; or a floating vessel may, by its rise and fall, communicate motion to machines.

Falling water; for machines employed to utilise, see Water Wheels, Section 90; Turbines, Water-pressure Engines, &c., Section 93.

Descending weights must first of course be raised, absorbing as much power in raising as they give out in falling, neglecting friction. Clockwork; water; or compression of a spring (see Section 80); multiplying pulleys (see Section 42), are the apparatus employed to utilise this form of energy.

Wave motion is too uncertain and erratic to be a practicable source of power. Rocking air-compressing chambers, rocking pumps, &c., have obtained some small measure of success.

Wind, windmills. See Section 95.

Expansion of air and gases. Ascending currents of hot air from a fire are used to drive a light screw motor, fan, &c. Hot-air engines, see Ryder's patent and numerous others, which depend upon alternate expansion and contraction of air by heating and cooling. Air compressed in an accumulator or reservoir is employed to give motion to multiplying pulleys or an air engine.

Expansion of liquids, other than water (by heat), into the gaseous form. Engines in which the fuel is burnt under pressure and the total products of combustion employed (with or without steam) to drive a motor.

Steam is in reality one of the last-mentioned sources of power; it is employed by direct pressure on a piston or ram (see Section 32); or to produce direct rotary motion (see Section 75); also in the jet pump, No. 801; or injector (see Section 45); or by direct pressure on a body of water contained in a closed vessel, as in the pulsometer, steam accumulator, &c.

Explosives are substances which, by application of flame, heat, percussion, &c., suddenly assume the gaseous form, thus increasing their bulk many hundred times, usually in a small fraction of a second of time. A second class comprise explosive mixtures of gases, such as hydrogen, and oxygen, carburetted hydrogen, and air. Some attempts have been made to employ explosive substances to drive engines in various ways, but with no permanent success. The second class of explosive mixtures of gases are largely employed in the gas engine, petroleum engine, and their varieties.

Fuels, hydrocarbons, &c., are employed to evaporate water into steam; to expand air or other gases, or convert liquids into gases; and also by vaporisation to supply gas for use in some forms of gas engine.

Section 61.—PUMPING ENGINES, TYPES OF.

VERTICAL ENGINES.

1114. **Vertical direct-acting,** with either ram pump, ram and piston pump, or piston only. See Section 56.

1115. **Slot and crank motion,** a variety of the last named. Of course any other kind of crank driving can be employed. See Section 10. The frame standards are frequently used as air vessels or valve chests.

1116. **Direct-acting ram pump,** with fly-wheel worked off crosshead pin.

1117. **Direct-acting,** with yoke crosshead; much used in the northern counties. The standards form air vessels and valve boxes, and they are made both of the piston and ram types.

1118. **Three-cylinder,** with yoke crosshead. Either the centre cylinder or the two side ones can be used as the steam motor cylinders, or the pumps.

HORIZONTAL ENGINES.

1119. **The ordinary direct-acting engine,** with either steam-moved or tappet valves, see Tangye's "Special," the "Coalbrookdale," and others, in which the slide valve is operated by pistons controlled by auxiliary tappet valves on the same principle as No. 1506.

1120. **Direct-acting,** with crosshead and guide bars between the cylinders.

1121. **Two modifications of No. 1120.**

1123. **Direct-acting,** with rocking lever valve motion; see the "Worthington" and other "Duplex" pumps, in which two engines are combined so that one works the valve of the other.

1122, 1124, & 1125. **Other forms of direct-acting engines.**

1126. **Horizontal compound direct-acting.** The high - pressure cylinder, low-pressure ditto, and receiver are side by side, and the air pump and main pump in line with the steam cylinders.

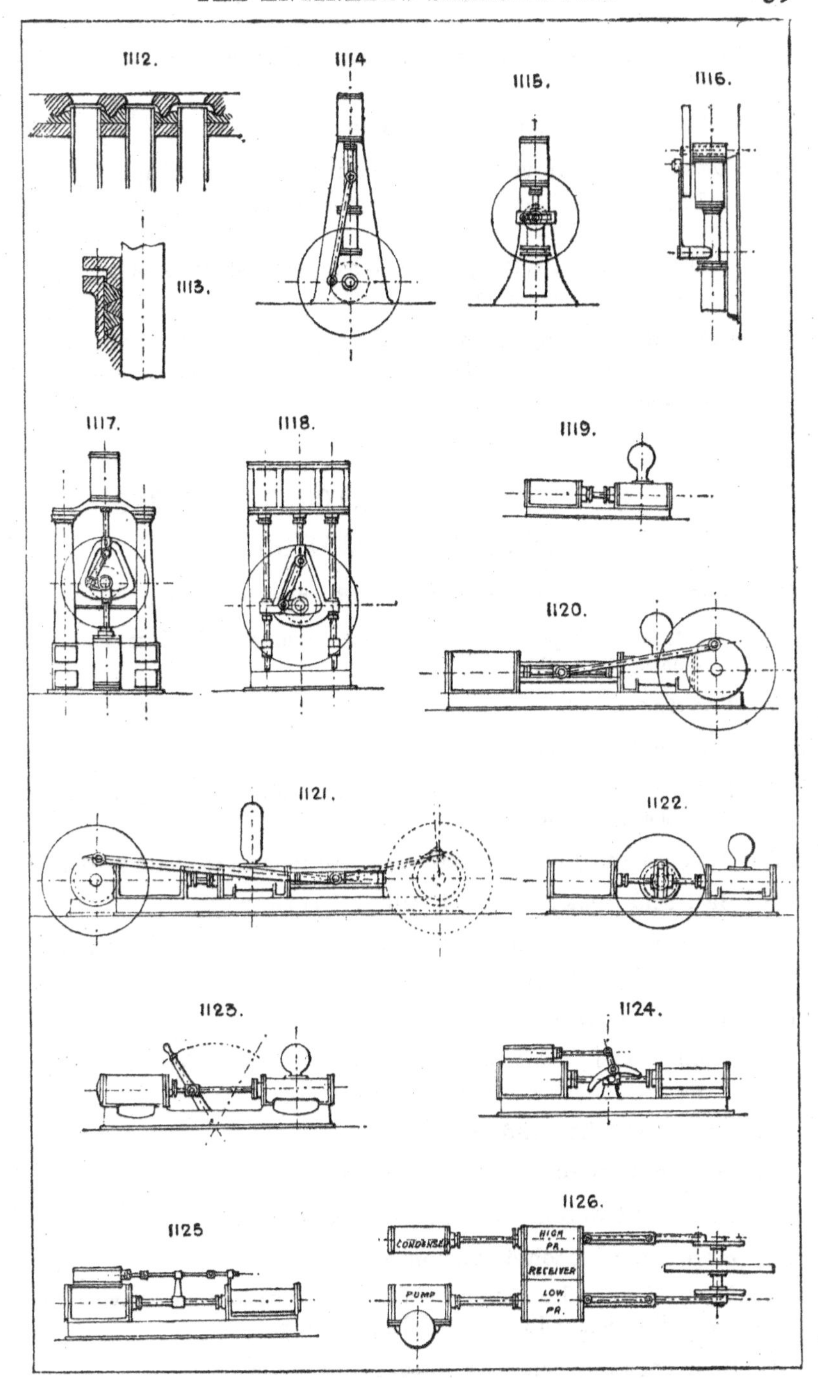
1112.
1113.
1114
1115.
1116.
1117.
1118.
1119.
1120.
1121.
1122.
1123.
1124.
1125
1126.
CONDENSER
HIGH PR.
RECEIVER
LOW PR.
PUMP

1127. **Horizontal pumping engine,** with yoked crossheads and crank in centre.

1128 & 1129. **Horizontal compound lever engines.**

1130. **Davey's patent vertical compound beam mining pump.**

The Cornish beam pumping engine is too well known to need illustration.

In mining pumps the pump rod has occasionally been made of iron pipe and employed as the rising main.

1131. **Geared pumping engine,** with steam cylinder and pump side by side; the speed of the steam piston is reduced on the pump by spur gearing.

Section 62.—PAWL AND RATCHET MOTIONS. INTERMITTENT MOTION.

1132. **The common ratchet-wheel and pawl,** or detent.

1133. **Ditto,** with compound pawls to check angular motion less than the pitch of the teeth.

1134. **Locking pawl.**

1135. **Strut-action pawl.**

1136. **Indiarubber ball pawl**; sometimes a solid roller is substituted for the indiarubber ball.

1137. **Reverse ratchets,** for continuous feed from an oscillating arm.

1138. **Ball and socket ratchet,** will work at an angle.

1139. **Pawl,** used with ordinary spur teeth, and something made reversible (see dotted lines), to drive the opposite way.

1140. **Ratchet bosses.**

1141. **Silent pawl**; the pawl is lifted out of gear while reversing by the motion of the toggle joint and lever.

1142. **Crown ratchet and pawl.**

1143. **Application of No. 1136** as a silent feed motion.

1144. **Click and detent continuous feed motion.**

1145. **Hare's foot ratchet motion** with detent.

1146. **Silent feed.** The jaw grips the rim of wheel when moving in one direction and runs loose the other way.

1147. **Reciprocating** into intermittent rotary motion.

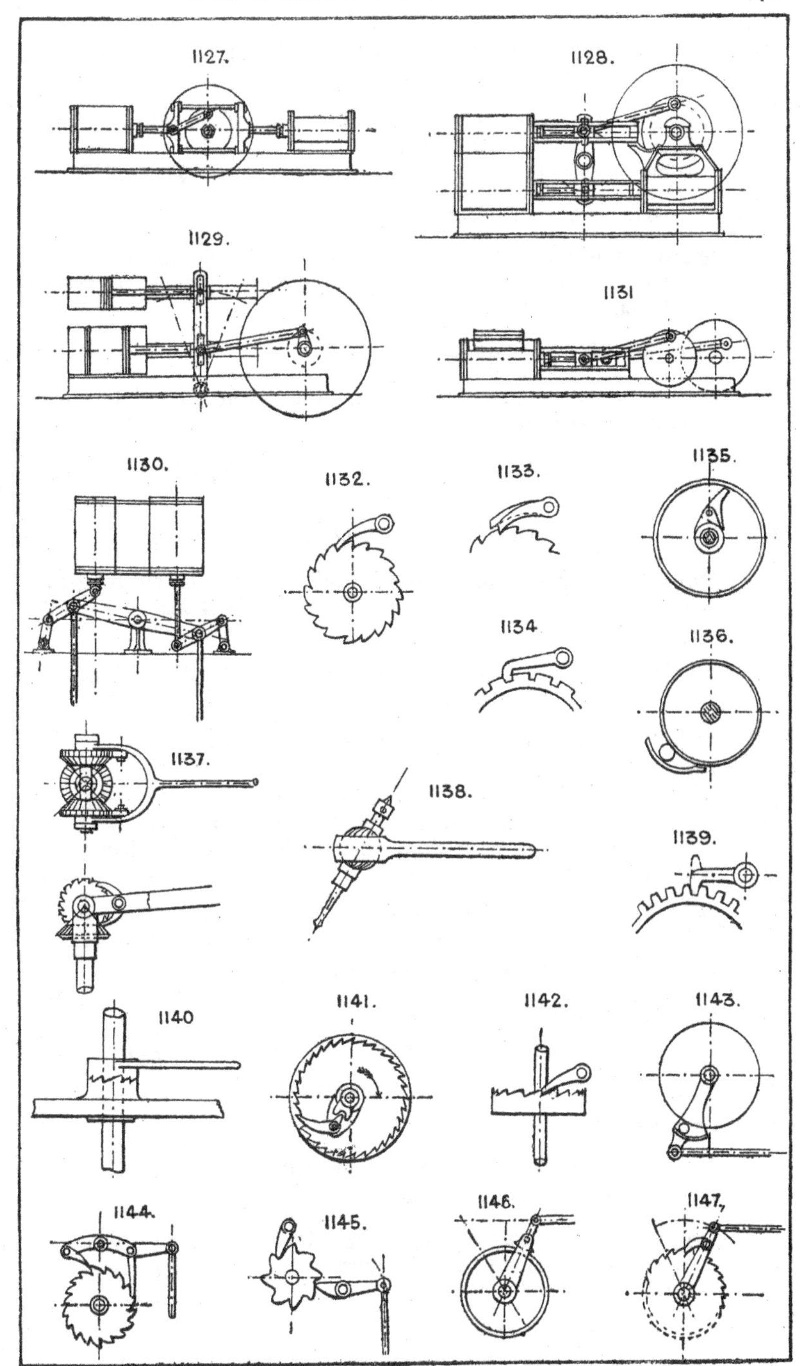
1127.
1128.
1129.
1131
1130.
1132.
1133.
1135.
1134
1136.
1137.
1138.
1139.
1140
1141.
1142.
1143.
1144.
1145.
1146.
1147.

1148. **Reciprocating circular motion** into intermittent circular ditto, Kaiser's patent.

1149. **Continuous circular motion** into intermittent ditto, Kaiser's patent. The wheel A is locked by the ring C while the finger B is out of gear, the ring then passing between the teeth of A.

1150. **Cam-ring intermittent feed motion.**

1151. **Modification of the last named**; in both the wheel is locked during the dead movement of the cam by the flange passing between the teeth.

1152. **Slot wheel and pin gear.**

1153. **Segment-wheel intermittent feed motion**; locked during the dead movement of driving wheel.

1154. **The pawl is lifted out of gear** at each revolution of the pin wheel A and the ratchet moved one or more teeth.

1155. **Double pawls and links** for continuous feed motion.

1156. **The cam** A is eccentric to the wheel B, and slips out of gear at any required point while the driving wheel makes a partial revolution.

1157. **Spring-pawl feed motion**; the large wheel with pawl attached drives the ratchet wheel.

1158. **Rocking lever and double pawls** for raising a rack.

1159. **Internal pawls,** dropping into gear by gravitation.

1160. **The pawl is lifted out of gear** by the act of putting the handle on the square end of shaft, the handle having a boss shaped so as to lift the pawl.

1161. **Star wheel and fixed pawl** for conveying intermittent motion to screw on revolving disc; used for boring-bars, slide rests, &c.

1162. **Pendulum and ratchet escapement.**

1163. **Cylinder escapement.**

1164. **Pendulum and double ratchet wheel escapement.**

1165. **Enlarged plan of cylinder escapement.**

1166. **Lever escapement.**

1167. **Double pawl and pin wheel escapement.**

1168. **Three-leg pendulum escapement.**

1169. **Self-sustaining ratchet motion.** Pulling the cord A throws the pawl out of gear by the straightening of the cord forcing back the bent pawl lever.

1170. **Verge escapement.**

1171. **Intermittent circular motion** by revolving pawl and detent.

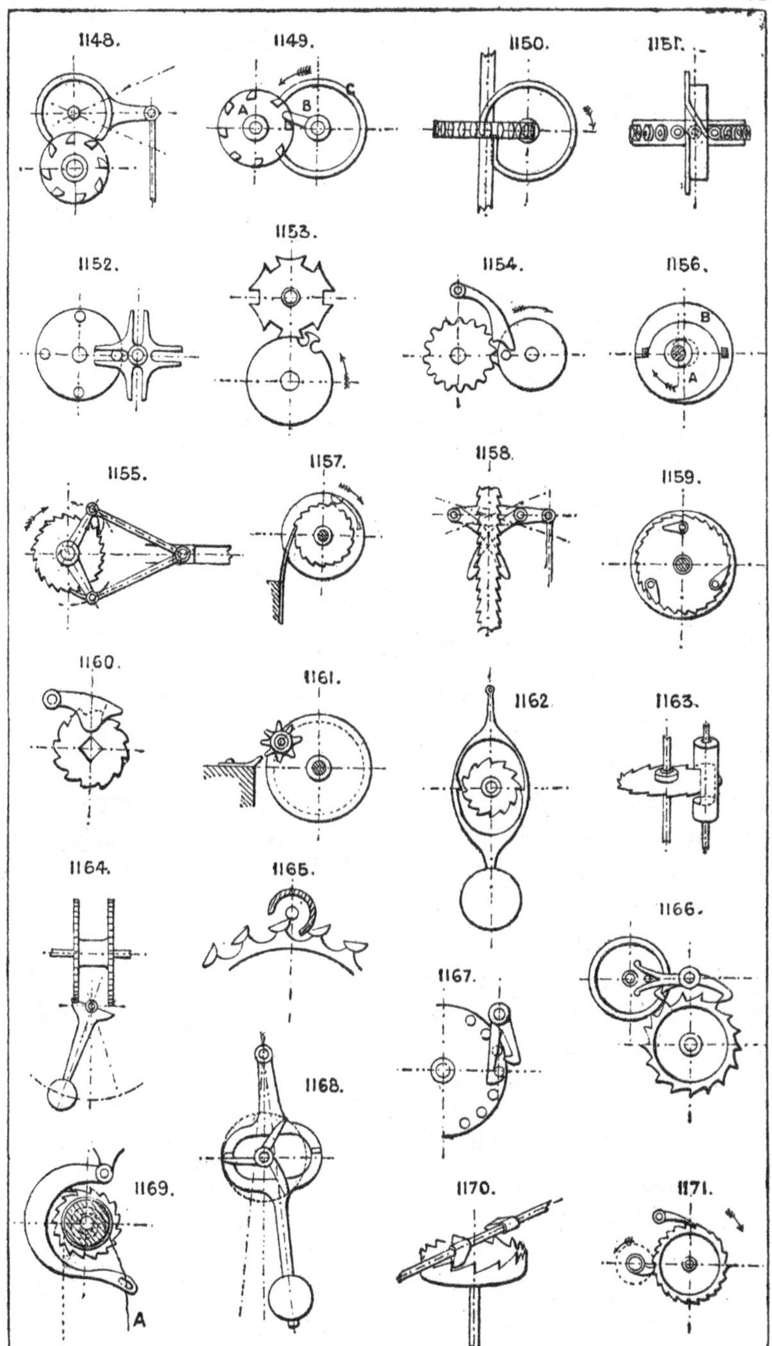
1148.
1149.
A
B
C
1150.
1151.
1152.
1153.
1154.
1156.
B
A
1155.
1157.
1158.
1159.
1160.
1161.
1162
1163.
1164.
1165.
1166.
1167.
1168.
1169.
A
1170.
1171.

1172. **Spanner ratchet**; a simple spanner having a pin near one end of one of the jaws which slips into the teeth of the rachet wheel.

1173. **V pawl**; operates by wedging itself between the V flanges.

1174. **Gravity pawls and ratchet wheel.**

1175. **Ratchet wheel,** used to govern the striking gear of a clock.

1176. **The pawl is hinged to the jointed end of the lever,** and is pulled out of gear by the return movement of the rod—silent feed.

1177. **Pawl and rack.**

1178. **Roller and inclined segmental recess** for silent feed motion.

1179. **Gripping pawls and ring** for silent feed motion.
Snail ratchet, No. 725.

Section 63.—PRESSING.

The ordinary Screw press and Hydraulic press are well-known machines.

1180. **Rack and screw press.**

1181. **Power press or stamp,** with double crank movement worked from below.

1182. **Dick's anti-friction press,** with rolling contacts throughout.

1183. **Hydraulic press,** with dies for lead pipe making; a similar press is used for making earthenware drain and flue pipes, the material being forced out of an annular orifice.

1184. **Wedge press.**

1185. **Ster-hydraulic press**; a strand or rope is wound upon a barrel inside the cylinder, thus displacing the water and raising the ram.

1186. **Screw fly press.**

1187. **Combined screw and hydraulic press.** The screw is worked down by hand until the pressure becomes too great for hand power, when the pressing is finished by the hydraulic ram.

1188. **Revolving dies.**

1189. **The "Boomer" double-screw toggle press,** with increasing pressure as the press follower descends.

1190. **Revolving toggle press,** with similar capabilities but more restricted movement.

1191. **Sector and link press** for increasing pressure.

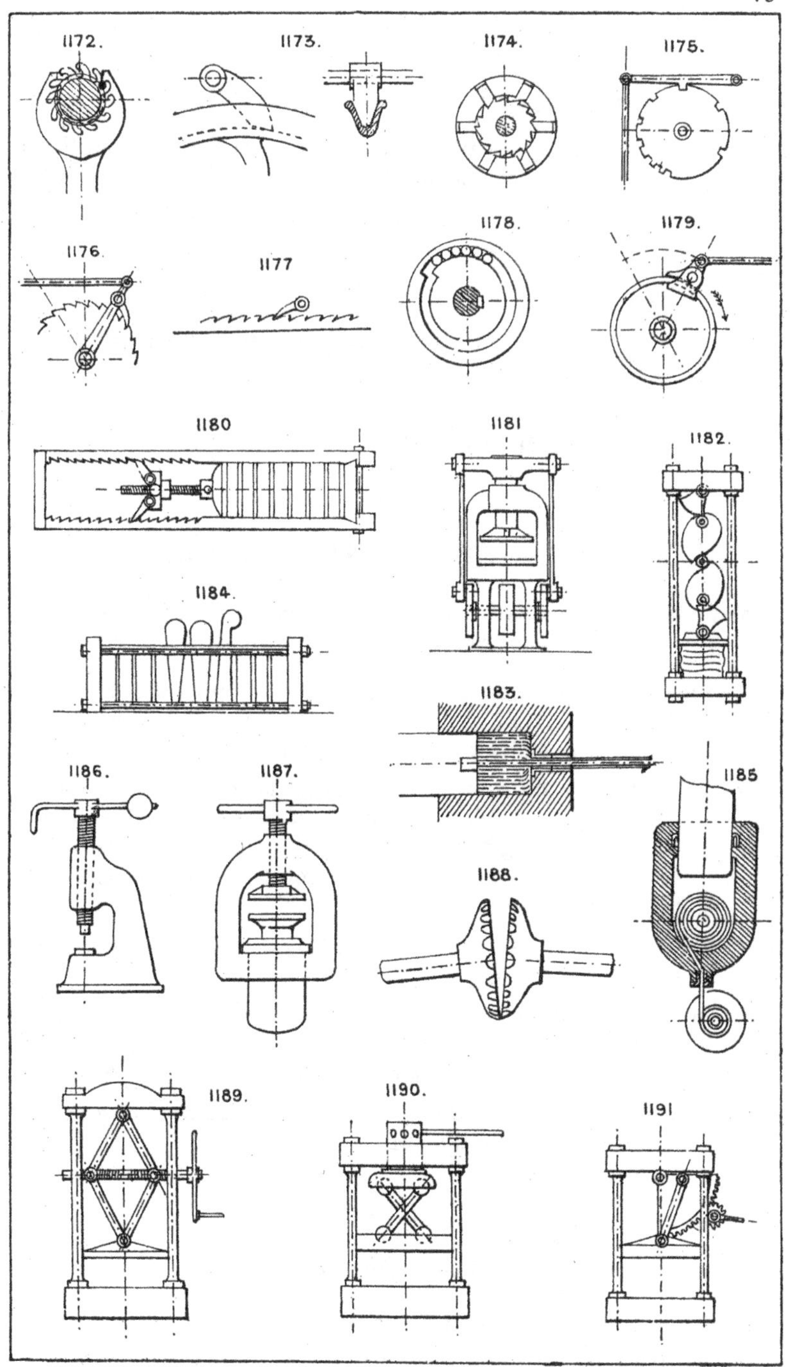
1172.
1173.
1174.
1175.
1176.
1177
1178.
1179.
1180
1181
1182.
1184.
1183.
1186.
1187.
1185
1188.
1189.
1190.
1191

1192. **Press dies,** with sliding plate for discharging.

1193. **Screw and toggle press;** a modification of No. 1189.

1194. **Double ram hydraulic press** for two pressures; the small ram is employed to give the first pressure, the large ram then finishes the pressing.

See also Section 13.

Section 64.—POWER AND SPEED, CONTRIVANCES TO VARY.

(See Mechanical Powers, Section 53.)

Section 65.—QUICK RETURN MOTIONS.

Employed for machines having a slow movement one way, and a quicker return movement.

1195. **Slot lever and crank motion;** gives a varying speed, quickest when the crank pin is at the bottom centre, and slowest when at the top, with a slight pause at each end of stroke.

1196. **Whitworth's motion.** The pin A in the wheel B travels eccentrically in the crank disc, which is eccentric to the fixed boss on which the driving wheel B runs, so that the radius of the driving motion of the pin varies as it revolves and it travels up and down the slot in the disc.

1197. **By two belts,** one open and one crossed, and driving drums of different diameters.

1198. **By two belts,** one open and one crossed, but driven by the same drum. The middle pulley is loose, the left hand pulley is fast to the spur wheel, and the right hand pulley to the spur pinion.

Segment gear, gearing alternately with the internal ring and the central pinion. See No. 724.

See Section 74.

Section 66.—ROPE GEARING.

1199. **V-grooved pulley rim** for round rope.

1200. **Multiple pulley rim,** used for rope driving in mills, &c.

1201. **Single V-grooved pulley** for hand ropes.

1202. **Pulley for wire rope** transmission, with wood bedding in the groove.

1203, 1204, & 1205. **Clip pulleys,** which grip the rope by its own tension.

Ordinary round grooved pulley. See Section 71, No. 1241.

Rope grip pulley, with snugs to wedge the rope and prevent slipping. See Section 71, No. 1242.

Pit head sheave. Used for quick running wire ropes. The boss is usually split to allow of expansion in cooling, and the arms are of wrought iron. See Section 71, No. 1243.

1206. **Rope driving.**

1207. **Rope driving,** with tightening pulley and weight.

1208. **Rope grip pulleys,** for driving a vertical rope, the large pulley has a V groove into which the rope is pressed by the small pulley.

1209. **Jigger hoisting rope gear,** used for whip cranes, &c., instead of spur gearing.

Wire rope transmission. Endless wire ropes of small diameter are used running over large pulleys and driven at a high speed (usually 3000 to 4000 feet per minute). This kind of power may be carried considerable distances and over uneven ground, but it is not desirable to have horizontal angles in the direction of the rope.

Section 67.—RESERVOIRS OF POWER. ACCUMULATORS.

a. **The fly-wheel** or its equivalent.

b. **Springs.** See Section 80.

c. **Weights.**

d. **Air or gas compressed into a reservoir;** air vessel, bellows. See Section 7.

e. **Water raised into an elevated reservoir** or tank, or pumped into a loaded accumulator. Variable Pressure Accumulator, No. 1586.

f. **Electricity stored in accumulators.**

g. **Explosives.**

h. **Pendulum.** Sometimes used to accumulate power to be given out suddenly, as in punching.

Section 68.—RECIPROCATING AND CIRCULAR MOTION, CONVERTING ONE INTO THE OTHER.

(See Circular and Reciprocating Motion, Section 21).
(See Pawl and Ratchet Motions, Section 62).

Section 69.—RAISING AND LOWERING.

(1.) BY HAND POWER.

a. **The ordinary winch and crank handle.**

b. **Winch,** worked by an endless hand rope and wheel, similar to Nos. 1210, 1220.

1210. **Hand rope and barrel hoist.** In this machine gearing may be interposed between the hand rope wheel and rope drum to increase the power and reduce speed.

c. **Differential blocks** of various patterns (see Weston's, Pickering's, Moore's, &c.). See Section 31.

d. **By screw gear,** as in the ordinary screw jack. Sec. 78.

e. **Rack and pinion gear.** See No. 754.

f. **Worm and wheel gear.** See Section 84.

Note as to brake wheels; these should always be upon the *load* shaft so that the braking is not transmitted to the load through toothed gearing. Worm gear usually will not sustain a load without a brake wheel, unless there is an excess of friction which should not exist.

g. **Friction gear.** See Section 38.

(2.) BY POWER.

This may be applied to any of the above as follows:—

a or b. To the ordinary winch by either gearing (see Section 84), belts (see Section 3), or friction gear (see Section 38).

b. **By gripping** the endless rope between grip wheels (see 1208) the small wheel can be thrown into gear to grip the rope by a lever, cam or screw.

c. **Differential gear** may be driven from a shaft by belts or gearing. See Sections 84 and 3.

d. **Screw gear.** Ditto.

e. **Rack gear.** Ditto.

f. **Worm and wheel gear.** Ditto.

g. **Friction gear** is usually driven as No. 1211.

1211. **Where the barrel shaft has a slight horizontal movement,** so that, by the lever, it can be forced into gear with the friction pinion to raise the load, or into the brake block to sustain the load or lower it.

Grooved friction V gearing is also sometimes used. See No. 667.

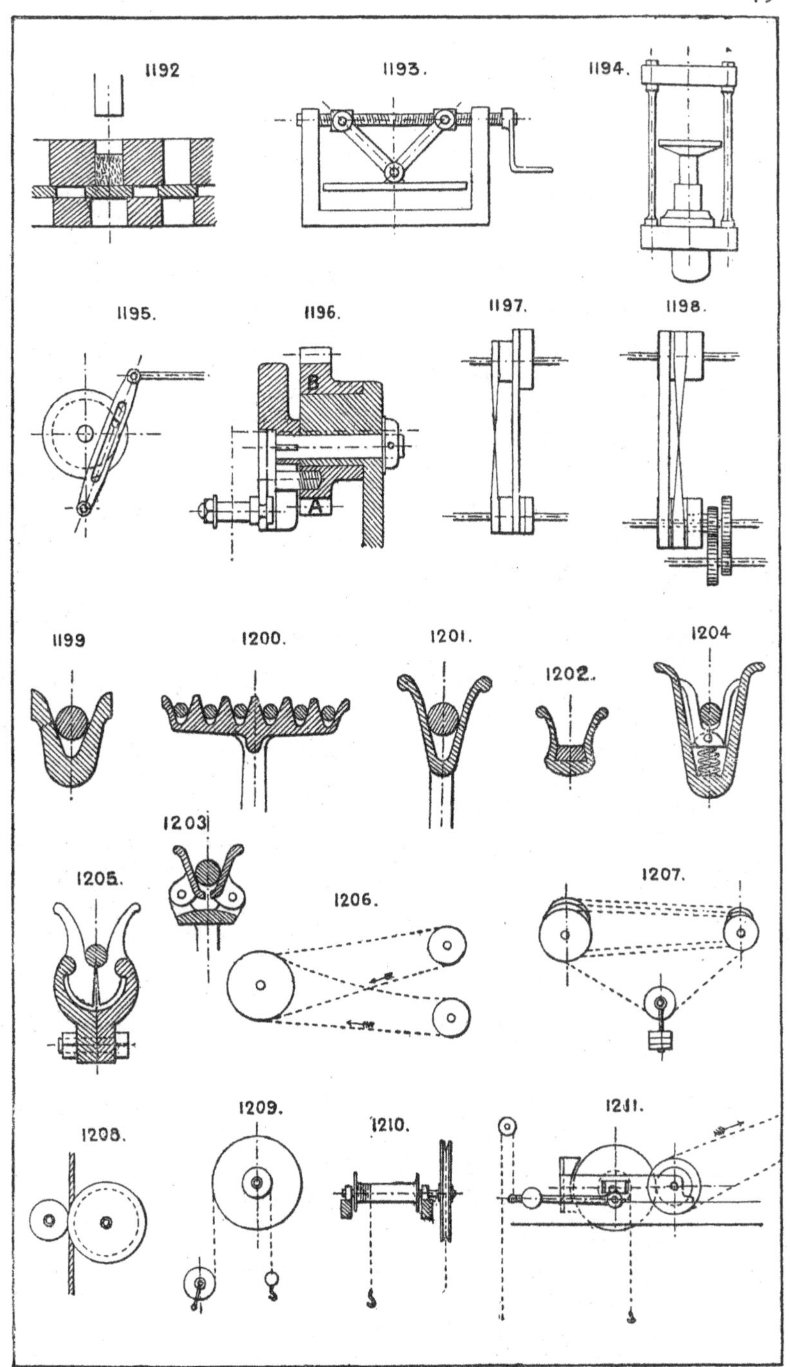
1192
1193.
1194.
1195.
1196.
B
A
1197.
1198.
1199
1200.
1201.
1202.
1204
1203
1205.
1206.
1207.
1208.
1209.
1210.
1211.

(3.) HYDRAULIC GEAR. See Section 42.

The direct-acting plan is simply a ram and cylinder, as in the hydraulic press, the ram being as long as the height of travel of the cage. For multiplying cylinder hydraulic gear, see Section 42.

Balancing the dead load of cage, &c. This is usually done by weights attached to the end of ropes running over overhead pulleys and fastened to the cage, as in No. 370, or by an auxiliary cylinder and ram of short stroke loaded to the required weight, and communicating with the lift cylinder by a pipe. See Section 20.

(4.) FOR LOWERING WEIGHTS ONLY.

a. **An hydraulic cylinder and piston** may be used, to which the cage is directly attached either above or below, the cage or platform being overbalanced by a counterbalance weight and rope (running over a pulley as No. 370), which is sufficient to raise it empty. The speed is controlled by a pass valve which allows the water to pass from one side of the piston to the other. See Section 5.

b. **An ordinary V wheel and brake wheel** may be used, the cage being overbalanced as last described; the motion is controlled solely by the brake. Or, an hydraulic brake cylinder may be used in connection with a rope or chain attached to the cage. See note to Section 5.

Other hoisting devices are:

Direct-acting steam or air cylinders, the piston rods being coupled direct to the cage.

Air vessels, on the principle of the gasometer, but of a height equal to the travel, and diameter proportional to the pressure of air employed.

1212. **Internal screw elevator.** The vertical shaft has a feather groove, and carries a double crosshead with a wheel at each end, which run on the spiral guides and raise the cage.

1213. **Screw elevator,** for ice, &c. Vertical creeper.

1214. **Travelling hoist,** with in and out motion and rope.

1215. **Steam digger and hoist.**

1216. **Hauling capstan.** The rope, which is payed on and off the barrel, "fleets" itself as it travels along the barrel owing to its conical shaped flanges.

1217. **Richmond's patent differential telescopic hydraulic lift.** The water under each piston is forced into the next cylinder above, so that the rams all travel upwards at proportional speeds, so as to reach the top of their stroke at the same time.

1218. **Self-sustaining gear.** The revolution of the pinion tends to lift the barrel and its brake wheel out of the brake; lowering is performed by relieving the brake wheel by a lever which raises it from the brake. Cherry's patent.

1219. **Belt hoist.** Worked by a loose vertical belt, which is tightened by the lever and pulley when required to hoist, and in lowering a load the belt friction acts as a brake.

1220. **Travelling hand hoist,** with endless rope.

1221. **Travelling cathead hoist.** The cathead can be run back with its load; the winch is sometimes fixed to the travelling beam and moves in and out with it.

1222. **Winding engine,** usual type for direct acting.

1223. **Geared winding engine.**

1224. **Steam winch, horizontal arrangement.**

1225. **Steam winch, diagonal arrangement.**

1226. **Steam winch, horizontal worm-gear plan.**

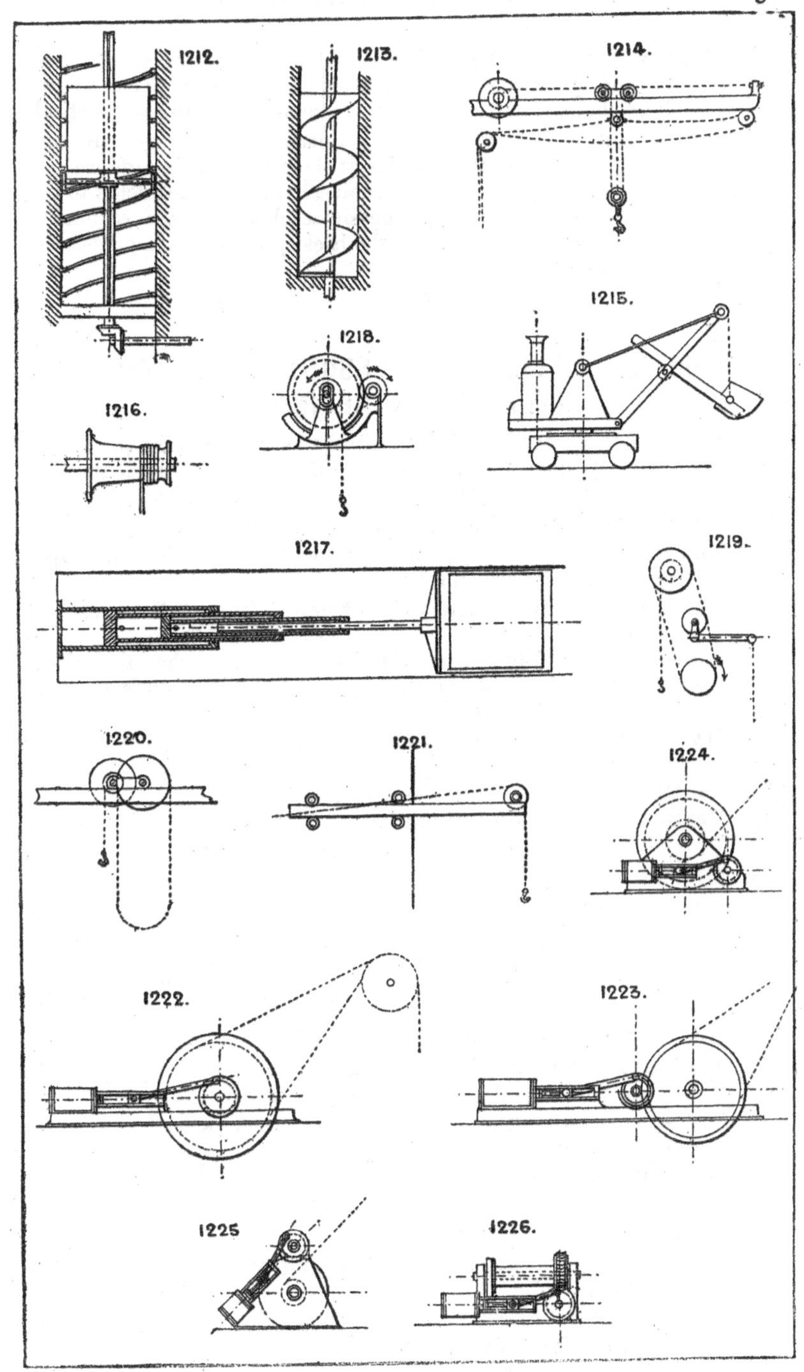
1212.
1213.
1214.
1215.
1216.
1217.
1218.
1219.
1220.
1221.
1222.
1223.
1224.
1225
1226.

1227. **Continuous lift,** for parcels, &c. Has a number of small cages, boxes, or platforms suspended from horizontal pivots in two endless chains; the cages are guided so as always to hang vertical.

1228. **Continuous lift,** for passengers. Sometimes the cages are suspended from two endless chains at their tops, as last described; or sometimes from two endless chains, but with attachments to the cages at corners diagonally opposite each other; and sometimes from a single endless chain at the back, provided with guides, &c.

1229. **Continuous barrel hoist.**

See also Cranes, Section 18.

Section 70.—RELIEVING PRESSURE ON BEARINGS. ANTI-FRICTION BEARINGS.

1230. **The pivots of two rollers or shafts** bear against the inside of a stiff ring producing rolling contact, but the rollers or shafts must run in the same direction. Used for roller mills, &c.

1231. **The same device, but for three rollers** or shafts.

1232. **The shaft is guided vertically** and its weight is borne by a large roller with small pivots.

1233. **The shaft runs in the V between two rollers** as last described.

1234. **Roller or ball bearing.** The friction is least when the rollers have end pivots to run in loose rings so that the rollers are kept apart and do not rub each other in revolving.

1235. **Hydraulic bearing,** the shaft being sustained by water (or preferably oil) pressure.

1236. **Vertical shaft,** with cone rollers.

1237. **Vertical shaft,** with ball bearing.

1238. **Vertical shaft,** flanged and coned for cone rollers.

1239. **Ordinary swivelling castor.**

1240. **Ball castor.**

Section 71.—ROPE, BELT, AND CHAIN PULLEYS.

(See also Section 66.)

1241. **Pulley for round rope** without any grip.

1242. **Round grooved pulley** for round rope with gripping snugs.

1243. **V pulley for round rope**; pithead pulley.

Pulley for wire rope transmission, high speed with wood bedding. See Section 66, No. 1202.

Multiple rope gripping pulley for rope driving. See Section 66, No. 1200.

1244. **Belt pulley,** flat face.

1245. **Belt pulley,** crown face. The rounding tends to keep the belt from running off.

1246. **Flanged belt pulley.**

1247. **Speed cone for belt.**

1248. **Round grooved chain pulley.**

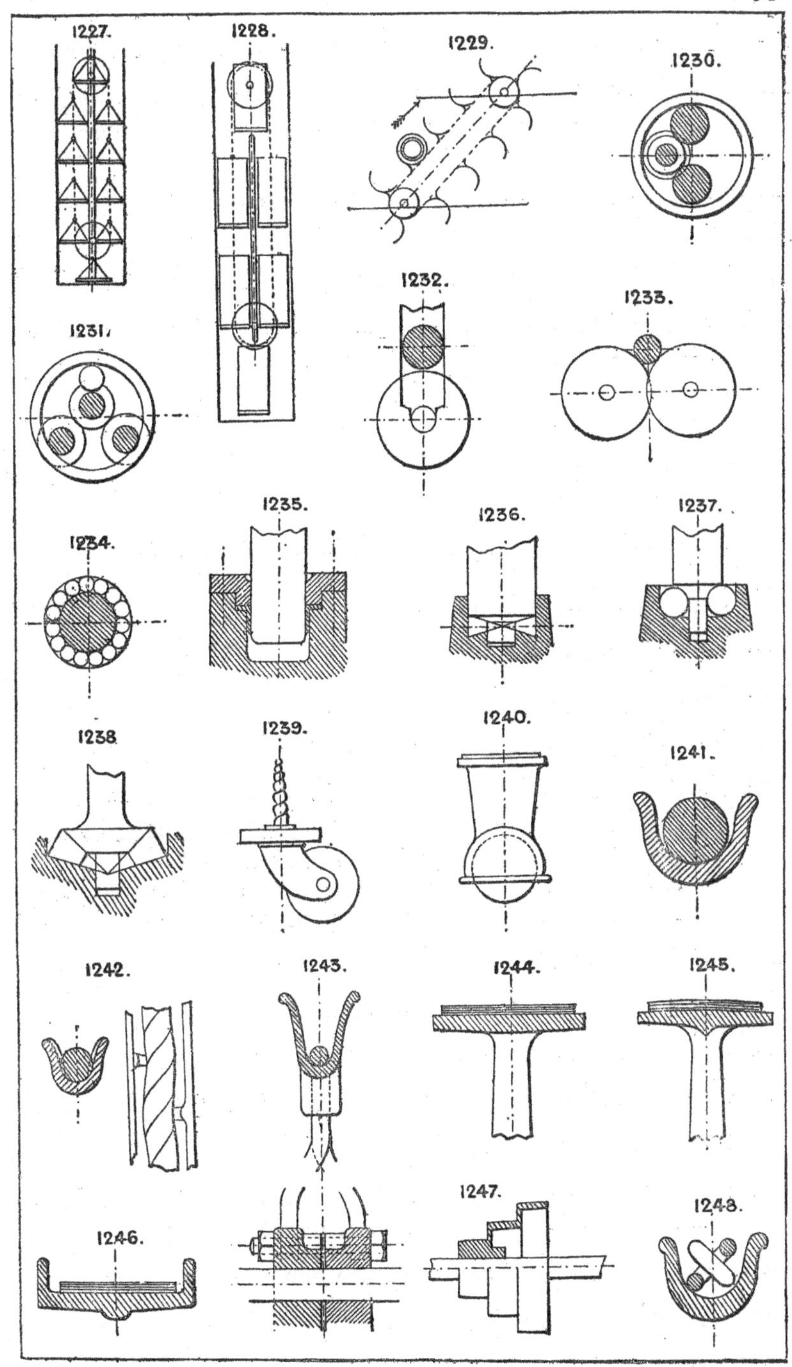
1227.
1228.
1229.
1230.
1231.
1232.
1233.
1234.
1235.
1236.
1237.
1238
1239.
1240.
1241.
1242.
1243.
1244.
1245.
1246.
1247.
1248.

1249. **Double grooved chain pulley**; prevents the chain twisting.

1250. **Pitched chain snug pulley or sprocket pulley.** The pitch of snugs should be slightly longer than the chain, to allow for wear and stretching of the chain.

1251. **Chain sprocket wheel,** for long link chain at slow speeds.

1252. **Sprocket wheel,** for long flat link pitched chains.

1253 & 1254. **Sections of rim showing single and double links.**

1255. **Sprocket wheel,** for Ewart's patent pitched chains. See Chains &c., Section 11.

Section 72.—RIDDLING AND SCREENING.

1256. **Square mesh wire gauze.**

1257. **Perforated plate.**

1258. **Parallel bars or wires.**

1259. **Hexagon or triangular mesh wire work.**

1260. **Slit and square hole perforations,** used for seeds, &c.

A form of variable mesh is manufactured by parallel series of diagonal bars jointed by pins to sliding cross-bars, so that the angle of the mesh bars can be altered and thus the spaces reduced or enlarged, on the same principle as No. 617.

1261. **Sloping screen.**

1262. **Cylindrical or slope reel screen.**

1263. **Cylindrical graduated screen or sizer.**

1264. **Rotary screen,** with rolling bevil gear motion. See No. 711.

1265. **Rotary horizontal screen.**

1266. **Shaking or jigging screen.** Sometimes supplied with a blast or aspirator to carry off the lighter particles.

1267. **Eccentric or angular barrel screen or mixer.**

1268. **Air blast sizing or graduating apparatus.**

1269. **Edison's magnetic sizing apparatus** for iron or steel particles.

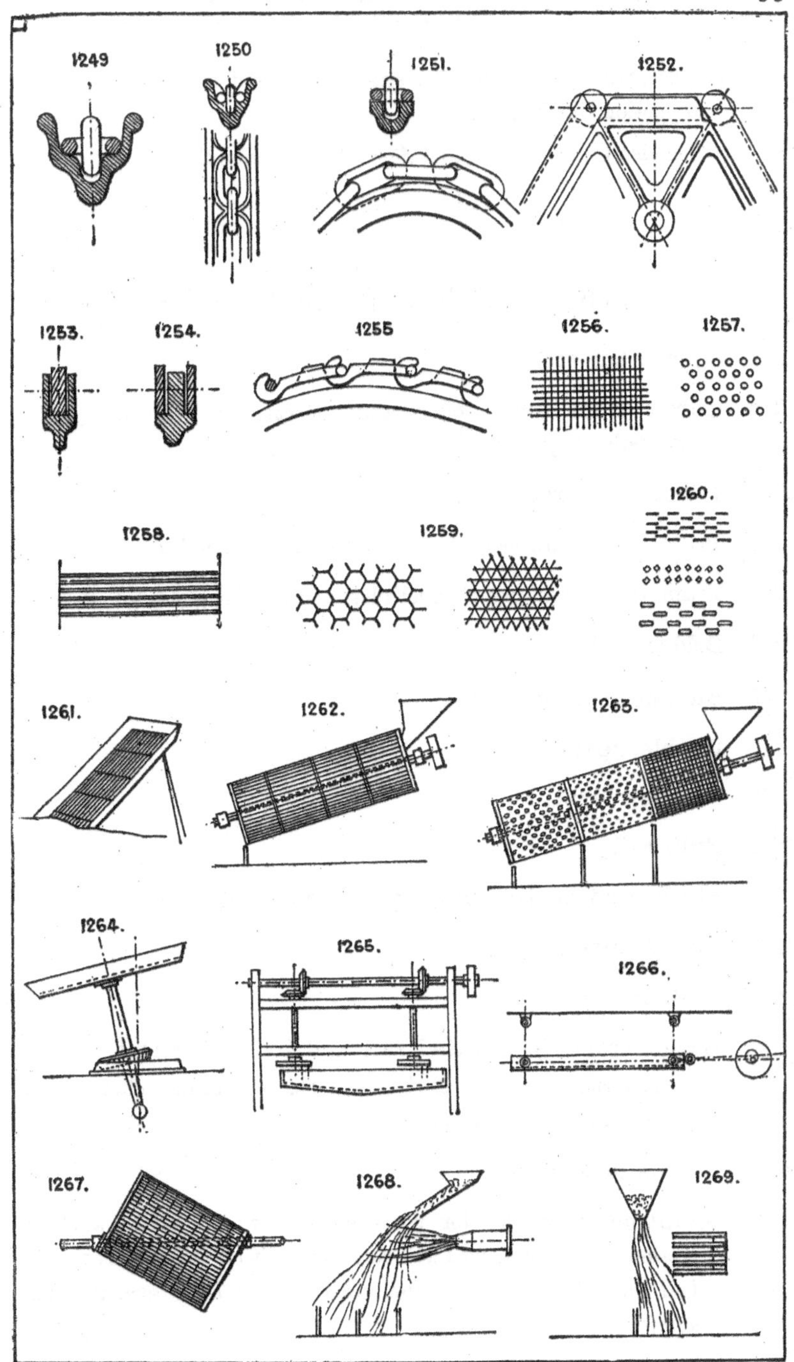
1249
1250
1251.
1252.
1253.
1254.
1255
1256.
1257.
1258.
1259.
1260.
1261.
1262.
1263.
1264.
1265.
1266.
1267.
1268.
1269.

1270. **Graduating or sizing screens**, either fixed as shown or kept in motion like No. 1266.

See also Concentrating and Separating, Section 26.

Section 73.—RAIL AND TRAM ROADS.

1271. **Square bar rails.**

1272. **L-iron tram road**; often made of cast-iron with the joints dove-tailed together.

1273. **T-iron tram road.**

1274. **Tram road**, with flanged plates for ordinary vehicles.

1275. **Tram road**, with one channel plate and one flat plate.

1276. **Bridge rail.**

1277. **Bulb-head flanged rail.**

1278. **Double headed rail.**

1279. **"Barlow" rail.**

1280. **Bulb rail.**

1281. **Flush grooved tramway rail.** See Nos. 1839–1841.

1282. **Rolled joist rail.**

1283. **Bulb-iron rail.**

1284. **Edge's patent perforated rail and toothed wheel.**

Many forms of combined chair and sleeper are manufactured in wrought iron and steel.

1285. **Left-hand switch.**

1286. **Shunting carriage**, for transverse shunting; carries a short section of the main road and runs across it on independent rails laid on a lower level; often used instead of a turntable for shunting.

1287. **Tramway switch.**

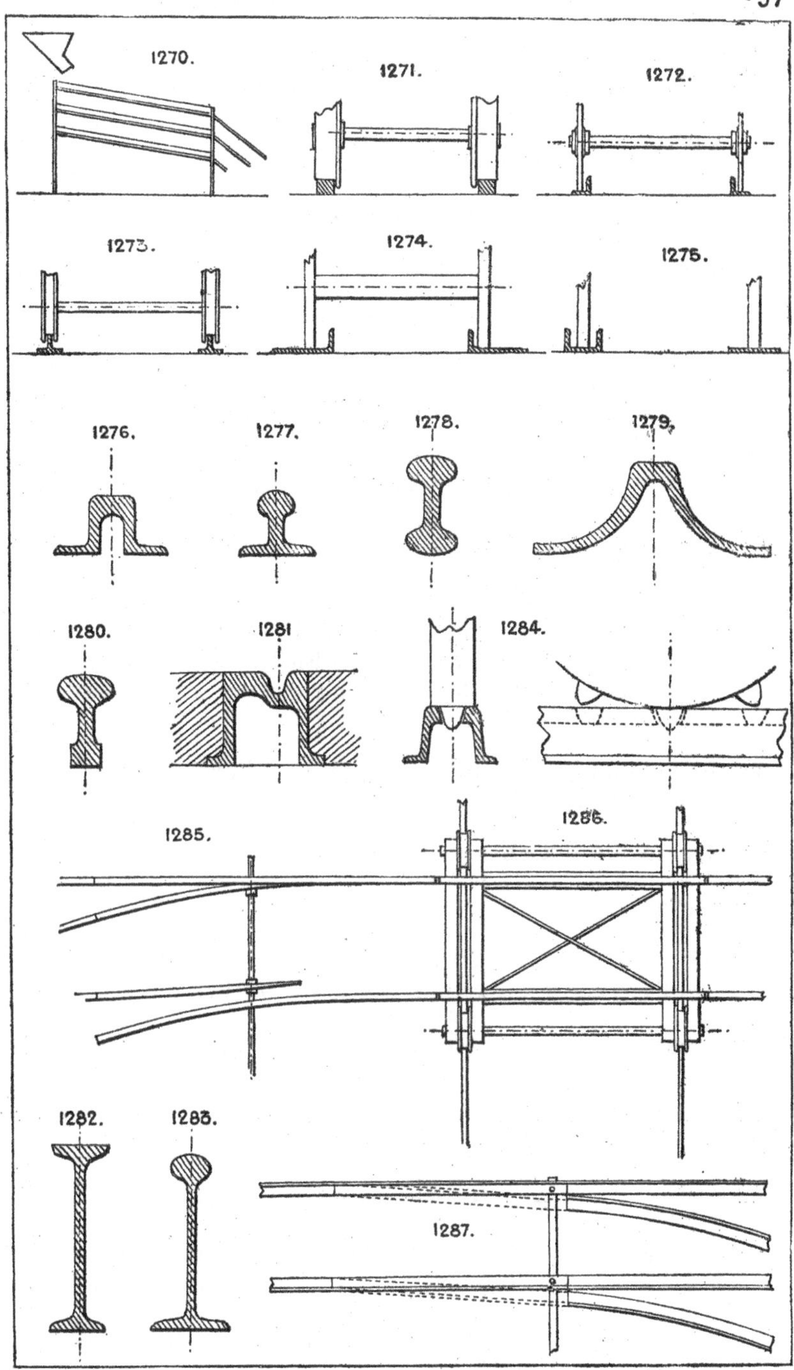
1270.
1271.
1272.
1273.
1274.
1275.
1276.
1277.
1278.
1279.
1280.
1281
1284.
1285.
1286.
1282.
1283.
1287.

1288. **Right and left-hand switch and crossing,** showing arrangement of guard-rails.

1289. **Flat bar** on edge rail.

See Section 99 for other sections of rails.

Section 74.—REVERSING GEAR.

For Reversing Gear of Steam Engines, see Section 79.

1290. **Reversible driving motion** by open and crossed belts, with two loose and one fast pulleys.

1291. **Reversible driving motion** by single belt, two fast pulleys and one loose ditto and bevil gear, one bevil pinion having a sleeve to which its fast pulley is keyed, the other bevil pinion being keyed to the shaft.

1292. **Reversible driving motion** by single belt, with quick and slow motions; a modification of the last.

1293. **By double clutch and bevil gear.**

1294. **Reversing friction cones or bevils.**

1295. **Three-wheel gear.** The driving wheel A can be put into gear either with the driven wheel C or idle wheel B.

1296. **Double clutch and spur gear** reversing motion, with idle wheel.

1297. **Reversing pinions,** as used on the ordinary screw-cutting lathe. There are many varieties of this gear in use.

1298. **Application of single belt gear to** No. 1296.

1299. **Self-reversing gear,** with one belt, two fast and one loose pulleys. The large spur wheel is driven from the bevil gear, and carries the weighted lever past the vertical position by a stop on the face-plate or disc, when it falls over and reverses the belt fork. See No. 1026.

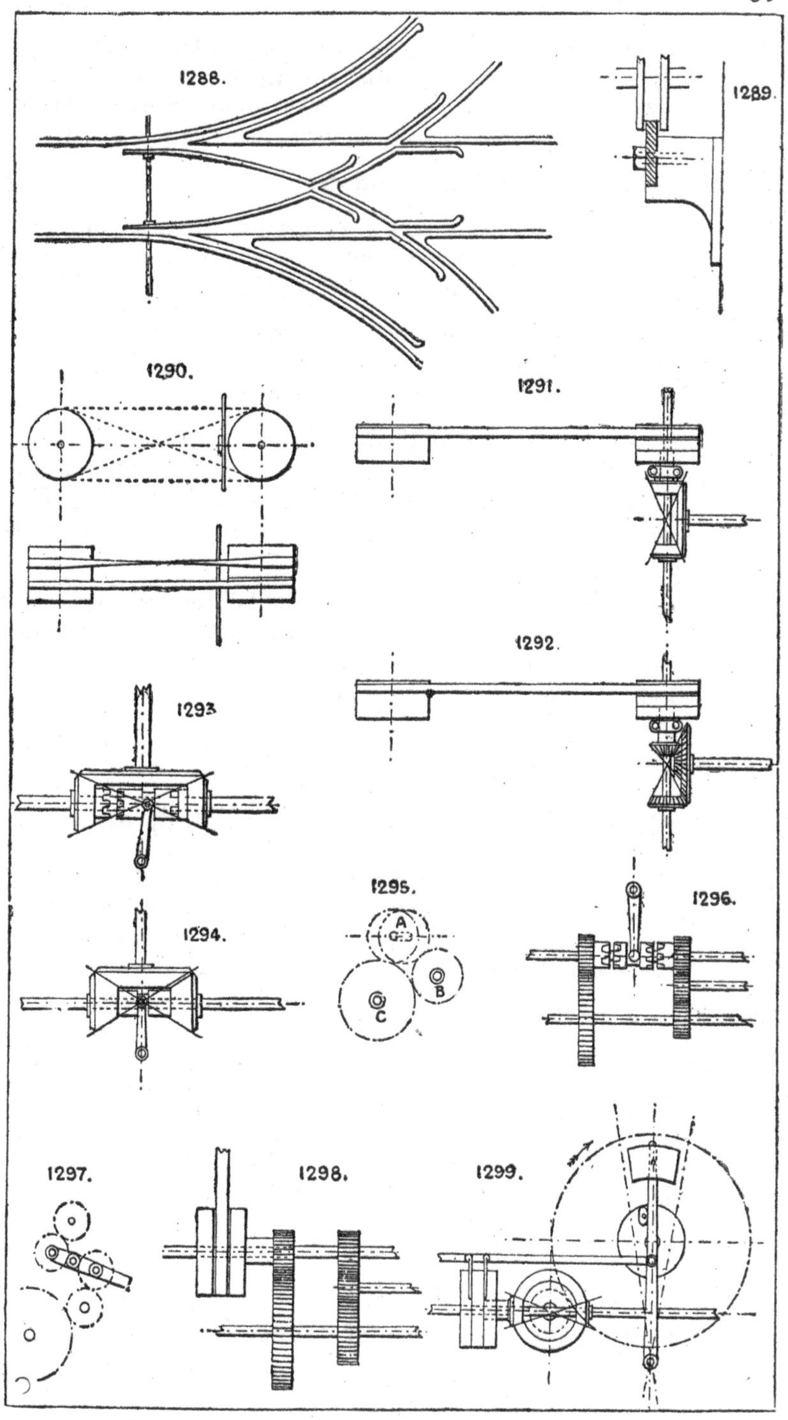
1288.
1289.
1290.
1291.
1292.
1293.
1294.
1295.
A
G
B
C
1296.
1297.
1298.
1299.

1300. **Self-reversing gear,** as applied to planing machines. The stops can be set at any required distance apart, to alter the length of travel of the machine bed. This plan requires a heavy table to carry the belt across the loose pulley to the other fast pulley.

1301. **Reversible belt-shifting hand gear.**

1302. **Right and left hand screw reversing traverse motion.** Each lever has a half nut, which can be put in gear with the screw to drive either way. See also No. 163.

1303. **Best form of fast and loose pulleys** for open and crossed belt reversing gear, as used in No. 1290; the fast pulley is rather larger in diameter than the two loose ones.

1304. **Single-belt reversing pulleys,** the reverse motion on the shaft being obtained by intercepting an idle wheel A between the epicycloidal wheel B and the shaft pinion C, the middle pulley being the loose one; the idle wheel is carried by a fixed bracket and pin.

NOTE.—Reversible motion can be obtained direct from any steam engine fitted with reversing motion. See Valve Motions, No. 1436, &c.

Segment Reversing Gear, No. 724.

Section 75.—ROTARY ENGINES, PUMPS, &c.

Nearly all rotary engines can be used either as motors, pumps, blowers, or meters, and most of the following typical devices have been applied to all four purposes. Most of them are reversible by simply reversing the direction of the motor fluid.

1305. **Disston's**; used as a pressure blower.

1306. **Root's, blower and pump.**

1307. **Root's.**

1308. **Mackenzie's**; may have one, two, or three vanes.

1309. **Gould's.**

1310. **Bagley and Sewall's.**

1311. **Greindl's rotary ump.**

1312, 1313, 1314, & 1315. **Varieties of intergeared piston rotary engines.**

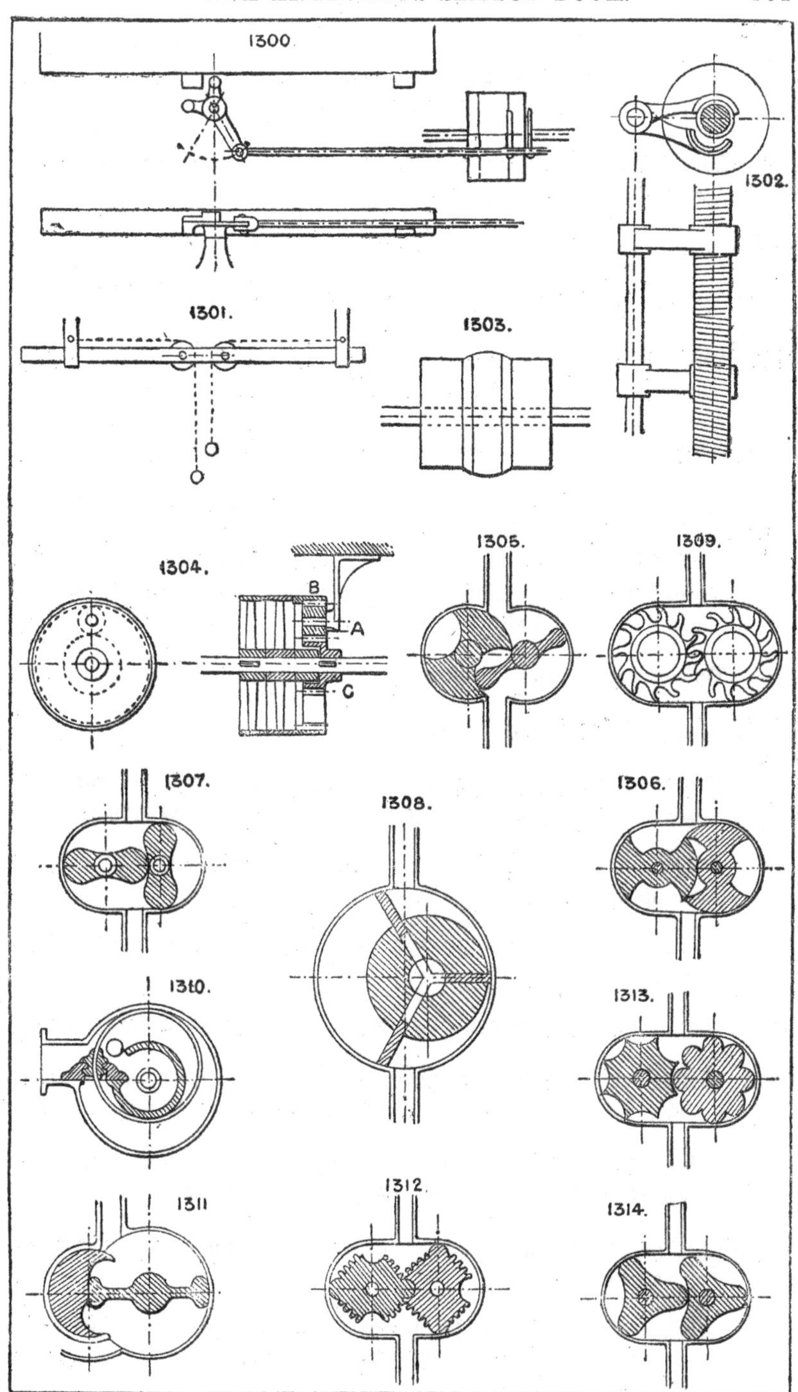
1300
1301.
1302.
1303.
1304.
B
A
C
1305.
1309.
1307.
1308.
1306.
1310.
1313.
1311
1312.
1314.

1316. **The small crescent-shaped piston** revolves three times to one revolution of the three-armed piston.

1317. **The hinged shutter** is thrown out of the way each time the revolving arm passes it.

1318. **Sliding shutter** and cam piston device.

1319 & 1320. **Varieties of the "Root" engine.**

1321. **The hinged vanes** are closed upon the revolving piston as they pass the flat side of the casing.

1322. **Has an eccentric piston and two hinged vanes.**

1323. **Eccentric piston and sliding diaphragm.**

1324. **Klein's motion.** The eccentric ring revolves in contact with the inner and outer casings, but is prevented from revolving on its axis by the fixed shutter and slot.

1325. **Baker's pressure blower.**

1326. **A modification of No. 1323.**

1327. **The eccentric ring** revolves on its centre, allowing the vanes to alternately project into the steam space as the wheel revolves.

1328. **Ivory's.** An eccentric cam and two sliding shutters, with a central steam inlet.

1329. **Mellor's** has a rocking vane oscillated by an eccentric piston carried round by a crank.

1330. **Eccentric piston and two sliding** vanes or steam stops.

1331. **Differential rotary engine,** with elliptical gear (see Section 34), or Stewart's differential gear (No. 554) may be employed.

The "Tower" spherical engine is a well-known form of rotary engine. See *Engineer*, August 10th, 1883.

1332. **An eccentric four-armed piston,** with four rolling stoppers or packings.

1333. **Mellor's patent pump** has a rocking vane or partition, with packing device which accommodates itself to the revolving oval piston.

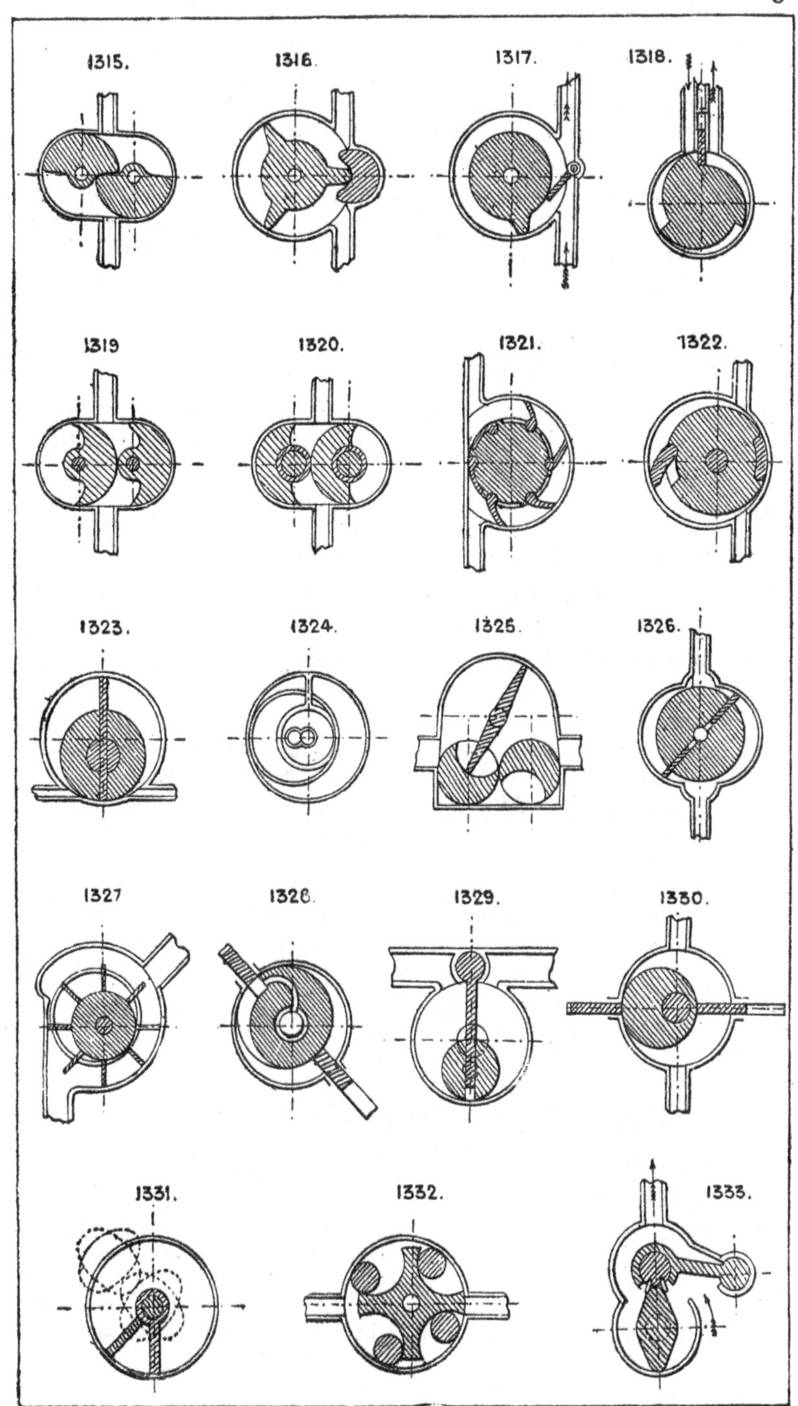
1315.
1316.
1317.
1318.
1319
1320.
1321.
1322.
1323.
1324.
1325.
1326.
1327
1328.
1329.
1330.
1331.
1332.
1333.

1334. **Bisschop's disc engine,** with three or four cylinders, single acting, whose rams press alternately on the edge of the disc.

1335. **Another form of disc engine,** in which partitions (rising and falling vertically) form the steam stops.

1336. **A modification of 1316.**

1337. **Rotary or centrifugal pump** or fan, numerous varieties of which are in use. Many of the later forms, as Blackman's and others, have the vanes fixed diagonally so as to propel the air at right angles to the plane of motion.

Section 76.—SHAFTING.

Employed to convey motion from a motor to various forms of driven machinery by gearing of various kinds. See Sections 3, 11, 38, 40, and 84.

Materials employed are:—Round, square, or polygonal wrought iron or steel bars, cast iron, wood, iron or steel tubes, planished round iron and steel bars, &c.

Stow's flexible shafting. See No. 442.

1338. **Longitudinal section of a cast-iron shaft.** These are sometimes made of a X section.

1339. **Wooden shaft** with end ferrules and iron end centres.

1340, 1341, & 1342. **End view and sections of ditto,** solid hexagonal and hollow circular.

1343, 1344, & 1345. **Arrangements of line shafting** in a machine-shop or factory, with or without overhead travelling crane.

1346. **Example of a line shaft,** showing bearings (see Section 46), couplings (see Section 16), pulleys (see Section 3), and gearing (see Sections 3, 84).

Shafts to be used as rollers are usually made hollow of wrought iron or other metal tube, tin plate, zinc plate, or sheet iron riveted; or sometimes, as No. 1342, of wood laggings fixed to solid polygonal blocks or centres, either continuous or in short pieces fixed at intervals.

See also next Section (77).

Section 77.—SPINDLES AND CENTRES.

1347. **Spindle** with sunk end bearings.

1348. **Spindle** with one sunk bearing and one collar.

1349. **Plain spindle** with two loose collars. Where pedestals with loose caps are used (see Section 46) the collars may be solid with the spindle, but with a long shaft the collars should be at one end only as shown, to allow of expansion. A wheel frequently occupies the place of one or both collars, and serves the same purpose.

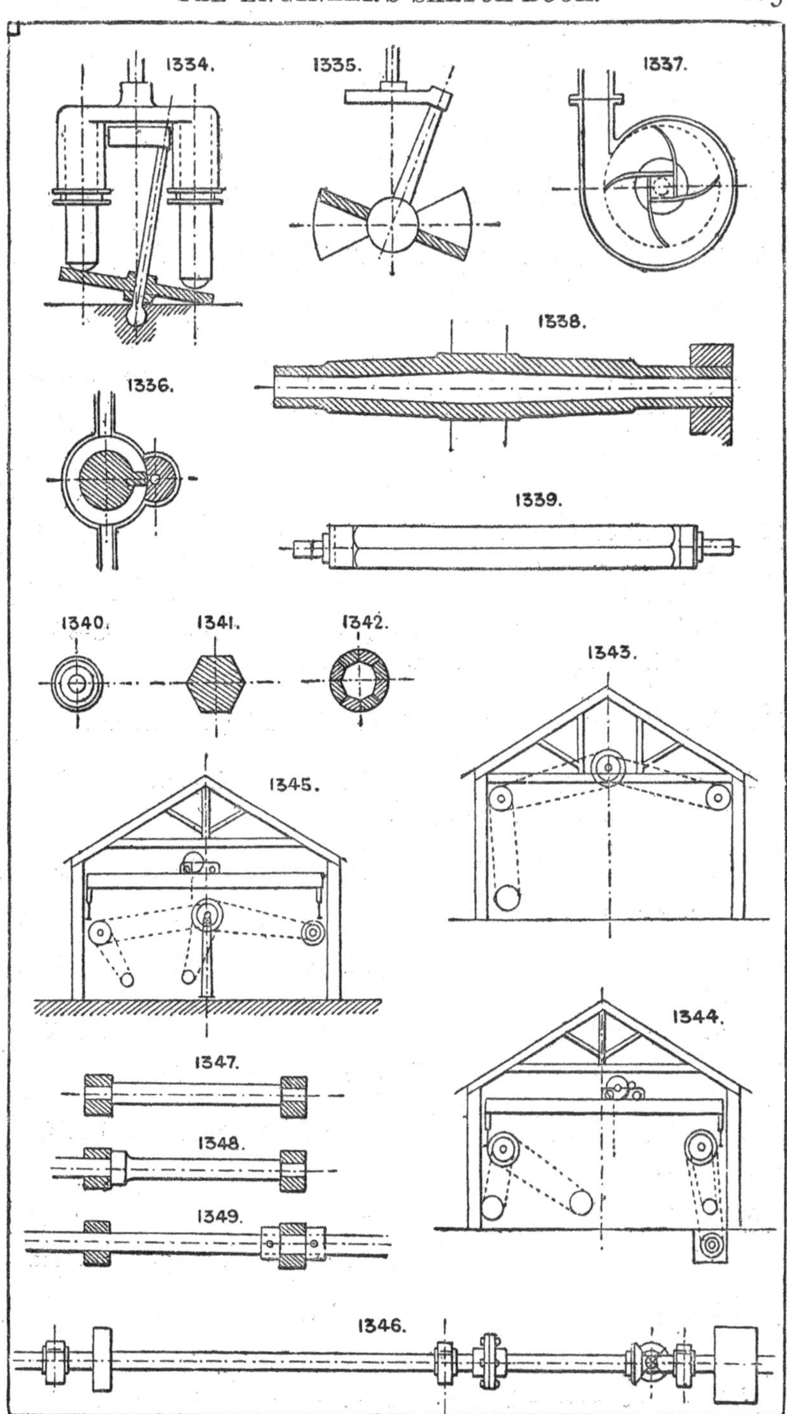
1334.
1335.
1337.
1338.
1336.
1339.
1340.
1341.
1342.
1343.
1345.
1344.
1347.
1348.
1349.
1346.

1350. **Coned centre,** fixed.

1351. **Collar centre pin or stud bolt,** fixed.

1352. **Coned centre** for roller or similar detail, driven in place.

1353. **Parallel centre** for roller or similar detail, keyed in place.

1354. **Square centre** for crane barrel, &c. See Nos. 634 and 635.

1355 & 1356. **Lathe headstock spindles,** solid or hollow; sometimes made with conical and sometimes with parallel necks.

1357. **Conical crane post.**

1358. **Conical cart axle.**

1359. **Universal centres,** employed to allow of a machine (or part of ditto), such as a drill, to be adjusted at any possible angle, the machine being fixed to one end of bar A.

1360. **Railway carriage axle.** For cranked axles, see Section 10.

1361. **Square neck centre bolt.** The square neck prevents the bolt turning and loosening the nut.

1362. **Coned and cottered crank pin or centre,** sometimes secured by a nut, as No. 1350.

1363. **Centre pin and bracket,** adjustable to various angles.

1364 & 1365. **Two methods of securing end of rod to any solid part of machine;** used for steam hammer heads.

1366. **Hollow post centre,** with water or steam channel to allow for swivelling.

1367. **Group of sleeve centres,** employed to allow of several pairs of lever or wheel motions being taken independently on a single shaft.

1368. **Ordinary centre pin,** with nut, washer, and split pin.

1369. **Ordinary centre pin,** with split pin and washer.

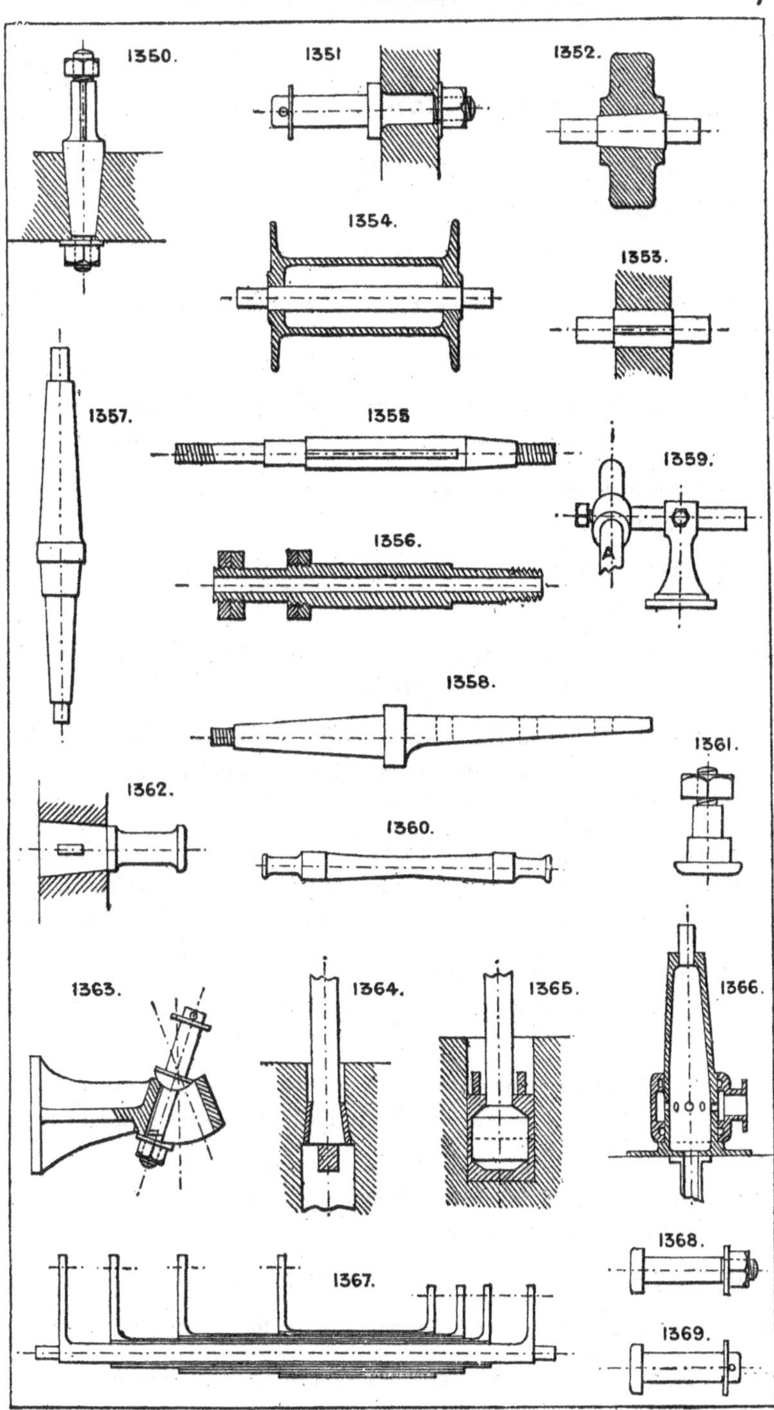
1350.
1351
1352.
1354.
1353.
1357.
1355
1359.
A
1356.
1358.
1361.
1362.
1360.
1363.
1364.
1365.
1366.
1368.
1367.
1369.

Section 78.—SCREW GEAR, BOLTS, &c.

1370. **Square thread screw.** Single, double, or multiple thread.

1371. **V thread screw.**

1372. **Enlarged section of V thread.**

1373. **Strongest thread** when the strain is always in one direction.

1374. **Round thread screw.**

1375. **Geared thread,** to be used with ordinary wheel teeth, the section of thread being that of a rack of the same pitch as wheel.

1376. **Earth screw,** screw pile, screw mooring, earth borer. See No. 530.

1377. **Fixed screw,** with hand wheel to revolve the nut, the screw having no rotation.

1378. **Conical screw;** used for chucks, &c. With two, three, or more sliding jaws chased to fit the conical thread.

1379 & 1380. **Differential screws.** One fixed, the other revolving, impart a motion equal to the difference of pitch of the two screws (see Section 31). See No. 1430.

1381. **Screw,** with half nut; the bearings of the screw being fixed act as a fulcrum for the motion of the half nut, which may be attached to any sliding device; employed for jaw chucks.

1382. **Screw and worm gear,** used for screw jacks, &c. The worm gears with a worm wheel having a central nut running on the main screw.

1383. **Mutilated screw and nut.** In one position the nut can slide on the screw, and a partial turn locks it. Used for instantaneous grip vices, &c.

1384. **Spiral worm** for three or four jaw chucks, expanding devices, &c. See Sections 28 and 36.

1385, 1386, & 1387. **Screw driver heads** for screws.

1388 & 1389. **Hexagon and square heads** for ordinary spanners.

1390. **Form of head** requiring special spanner or pointed bar.

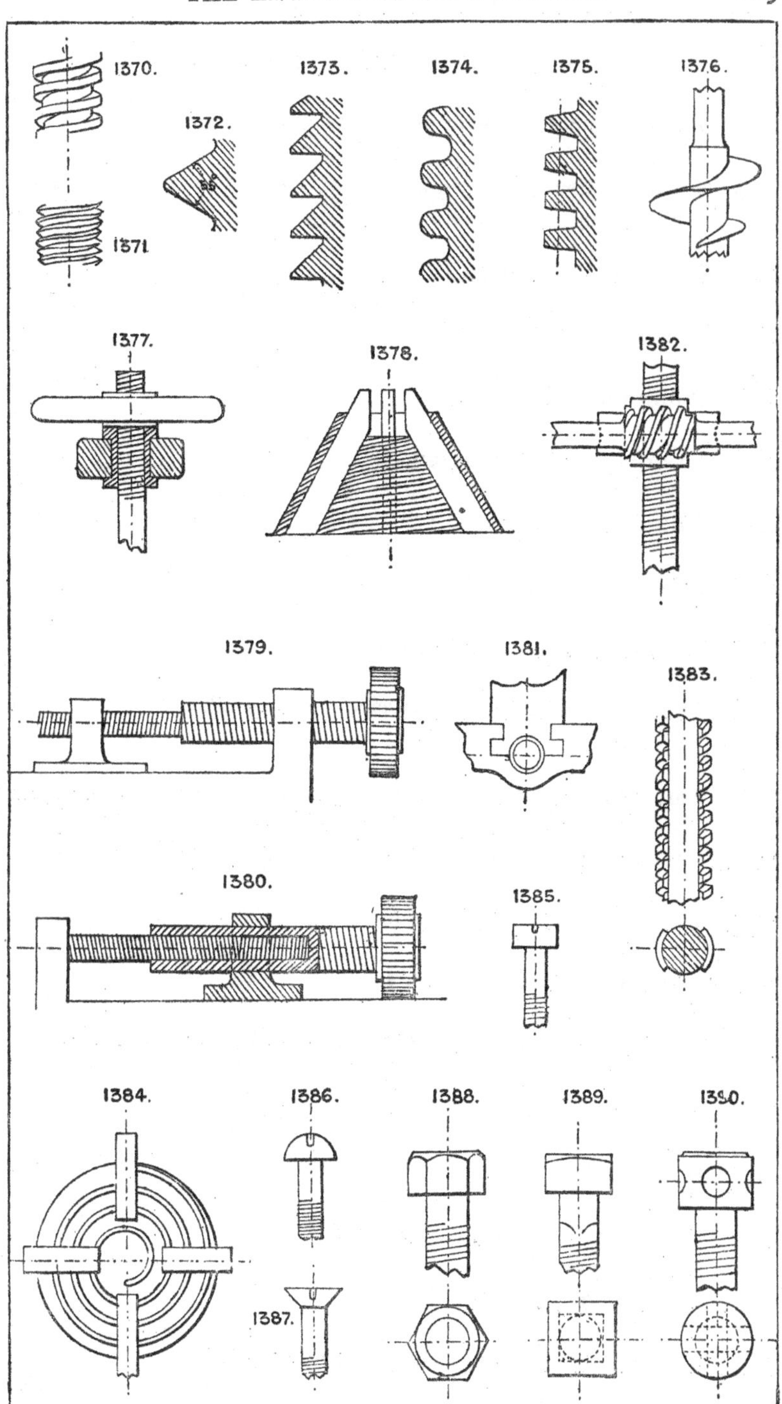
1370.
1371.
1372.
55°
1373.
1374.
1375.
1376.
1377.
1378.
1382.
1379.
1381.
1383.
1380.
1385.
1384.
1386.
1387.
1388.
1389.
1390.

1391. **Cylinder head bolt,** with drilled holes and special spanner.

1392. **Cylinder head,** but with flutes instead of holes for spanner

1393. **Cylinder head,** but with two flats cut on the head to suit an ordinary spanner.

1394. **Socket head,** to receive a second screw.

1395. **Eye bolt.**

1396. **Thumb screw.**

1397. **Thumb or shutter screw.**

1398. **Milled head screw.**

1399. **T head screw.**

1400. **Thumb or fly nut and screw.**

1401. **Hexagon collar stud** to receive a nut or other female screwed fixing.

1402. **Bolt head** for forked spanner, used for sunk or countersunk heads.

1403. **Hexagon head,** with solid washer or collar.

1404. **T head bolt** for T grooves in castings.

1405 & 1406. **Countersunk heads.**

1407. **Eye bolt,** with flat sides and straight eye for a pin or bolt.

1408. **Snap head.**

1409. **Hook bolt.**

1410 & 1411. **Lewis bolts, rag bolts.**

1412. **Cottered bolt.**

1413, 1414, & 1415. **Lewis bolts and key pieces.**

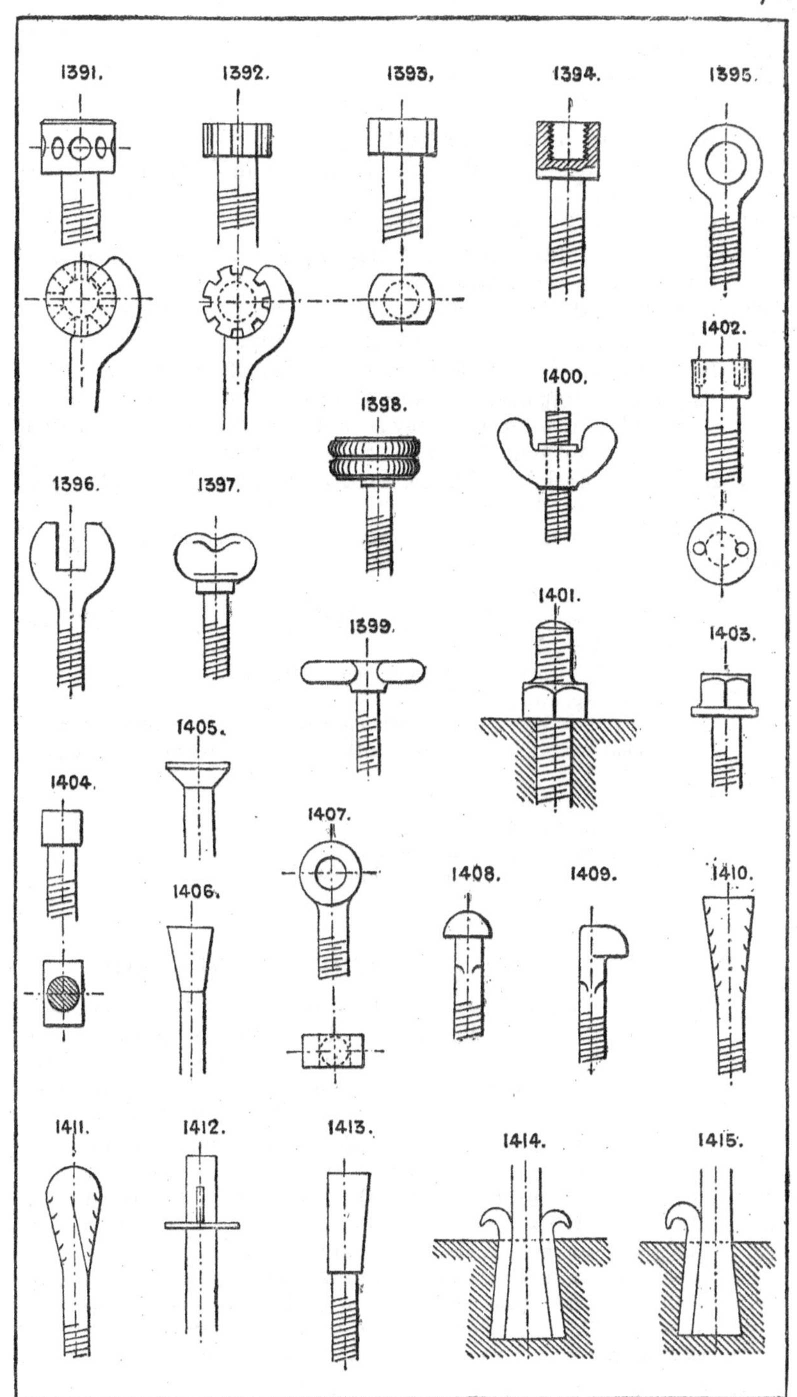
1391.
1392.
1393.
1394.
1395.
1402.
1400.
1398.
1396.
1397.
1401.
1399.
1403.
1405.
1404.
1407.
1408.
1409.
1410.
1406.
1411.
1412.
1413.
1414.
1415.

1416. **Collar stud.**

1417. **Split spring head bolt.**

1418. **Hook bolt.**

1419. **Solid head and collar bolt, bed bolt.**

1420 & 1421. **Heads for bolts** to slide and turn in T grooves of planing machines, &c.

1422. **Countersunk bed bolt.** Boiler stay.

1423, 1424, & 1425. **Methods of finishing screw heads** to prevent catching passing articles.

1426. **Screw head,** with cross dovetails to carry a key or screwing lever.

1427 & 1428. **Right and left-hand screw couplings** for tie rods, &c.

1429. **Ring coupling** for 2, 3, 4 or more rod ends for tie bracing.

1430. **Right and left-hand screw couplings** with halved ends to prevent the rods turning; may be made with one fine and one coarse thread for differential motion, or with right and left-hand threads.

1431. **Rifling,** as used in ordnance, &c., i.e. an internal multiple screw thread of very long pitch.

1432. **Screw spanner;** the weight prevents it working loose.

1433. **Belt screw.**

1434. **Gib cotter bolts.**

See also pipe couplings, Nos. 1071, 1072, 1073, 1074, 1075, and 1062, 1068, 1070.

For spiral worms and creepers, see Section 57.

Spiral pump, No. 1022.

Note that it is possible to construct a screw with an increasing or decreasing pitch, as is done with the screw propeller. See also No. 1378.

Double screw gear, No. 727.

Snail worm gear, No. 730.

Worm and crown gear, No. 733.

Worm and spiral gear, see Section 84.

Section 79.—SLIDE AND OTHER VALVE GEAR.

It would be neither easy nor useful, besides being beyond the scope of this work, to attempt to illustrate all the varieties of gear employed to work the valves of steam and other motor engines. I shall therefore only illustrate the more important types in general use, the details of which may be varied to suit individual cases.

1435. **Is the ordinary slide valve gear** with single eccentric for engines running always in one direction.

1436. **Ordinary link motion reversing gear** with two eccentrics; the link, having a shifting motion, is so arranged that either eccentric can be put into gear with the slide valve, the other eccentric running idle; or when in the mid position, as in sketch, both eccentrics run idle and the slide valve has no motion. By setting the link at intermediate positions the travel of the valve can be varied, and consequently the cut off also within certain limits.

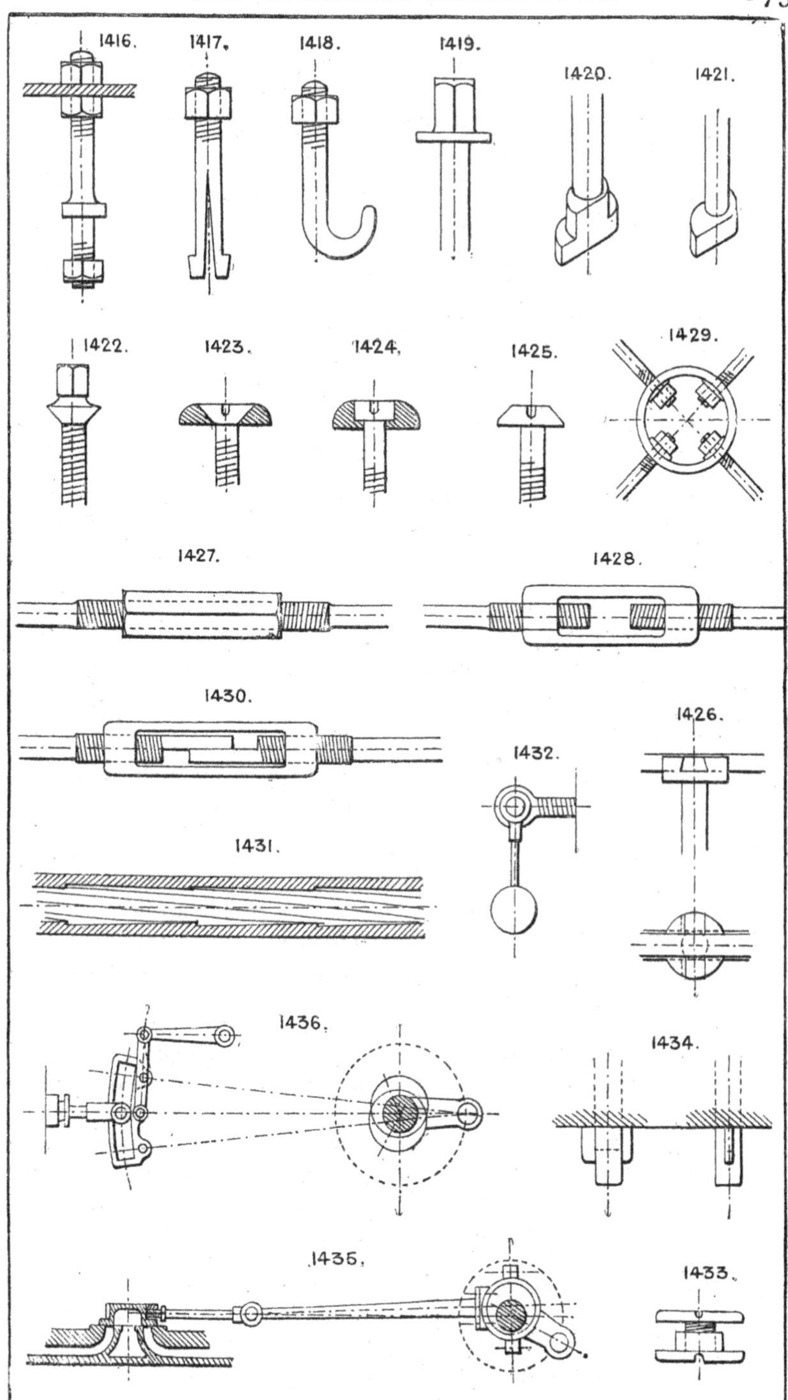
1416.
1417.
1418.
1419.
1420.
1421.
1422.
1423.
1424.
1425.
1429.
1427.
1428.
1430.
1426.
1432.
1431.
1436.
1434.
1435.
1433.

1437. **Nicholson's patent reversing gear,** without eccentrics. The drawing explains itself. This gear cannot be run in the intermediate positions, as a link motion to vary the cut off. This limits its usefulness to simple reversing only.

1438. **Automatic governor expansion** for single eccentric engine; the position of the connecting rod end in the swinging link is dependent on the governor, and thus also the travel of the slide valve.

1439. **Side shaft motion** for operating Cornish, Corliss, and spindle valves. The valves can be driven from this shaft by cams, eccentrics, or gearing.

1440. **Gab lever** for throwing the eccentric out of gear and thus stopping the engine.

1441. **Sector and link reversing motion** for oscillating engine; sometimes a shifting eccentric is used instead of link motion, as No. 1443.

1442. **Reversing sector link motion** for an oscillating engine; the valve is operated from the link, the angle of which is altered by the hand lever, there is therefore no lead to the valve.

1443. **Shifting eccentric and balance** sometimes used for reversing instead of double eccentrics and link; the loose eccentric is carried round in either direction by a stop piece on the shaft, fixed so as to give the correct lead both ways.

1444. **Murdoch's variable expansion gear** (see *Mechanical World*, September 29th, 1888) has one eccentric which operates a double arm lever, the outer end of which moves a sliding fulcrum along the valve rod lever, so that the leverage of the valve rod lever varies at different parts of the stroke. The sliding fulcrum is attached to a radius rod.

1445. **Proell's automatic expansion gear.** Shown applied to special double beat valves, but is sometimes applied to a special throttle valve, and is then applicable to any ordinary engine. The action of the governor alters the lap of the catches upon the ends of the valve levers, thus varying the time that the valves are kept open; the catches are centered on an oscillating T lever, operated from an eccentric on the main shaft.

1446. **Marshall's valve gear,** driven by one eccentric on crank shaft. The sector rocking centre is moved along the curved slot to reverse the engine, giving similar motion to the valve rod as in the case of the ordinary link reversing gear No. 1436.

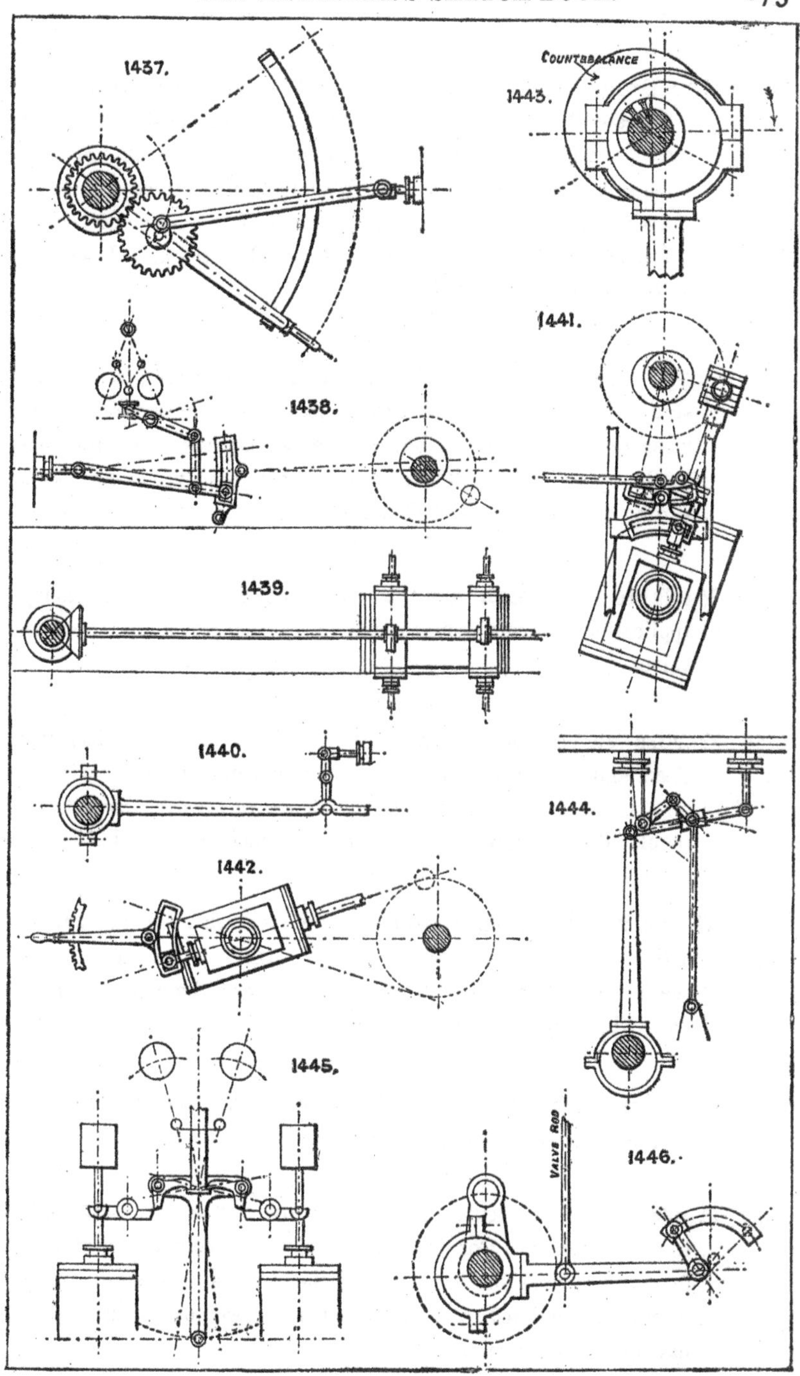
1437.
COUNTERBALANCE
1443.
1441.
1438.
1439.
1440.
1444.
1442.
1445.
VALVE ROD
1446.

1447. **The Bremmé valve gear** with single eccentric; the valve rod is operated by a lever and bent connecting rod from the end of the eccentric rod; the latter is constrained to move in an arc by a three-link attachment to a fixed bearing behind the eccentric rod and a movable one at the right-hand end of the horizontal link. To reverse; the arm and sector are turned to the dotted position by the worm and hand wheel shaft.

1448. **Joy's valve gear,** operated by a pin on the connecting rod. The slotted T lever is connected to the hand lever for reversing, and when reversed stands at the same angle from a vertical line but on the opposite side. The fulcrum of the valve rod lever has a sliding motion in the slot of the T lever.

1449. **Variable expansion gear** by hand power. There are many applications of this type used to vary the travel of a cut-off valve.

1450. **Corliss valve gear,** operated by a single eccentric, has two steam and two exhaust valves similar to No. 1642, worked from pins on a rocking wrist plate. The steam valves have trips, regulated by the governor, on a similar principle to No. 1445.

1451. **Crank shaft governor** with shifting eccentric: the centrifugal action of the weights, acting against springs, is used to revolve the inner eccentric so as to vary the throw of the main eccentric from which the slide valve is driven.

1452. **Another form:** in this also the throw of the eccentric is varied by the action of the governor ball.

1453. **Another form of automatic governor expansion trip gear** in connection with Cornish valves; a single eccentric operates the four valves, and the contacts of the catches and steam valve levers are regulated by the governor, the lower or exhaust valves having a constant motion.

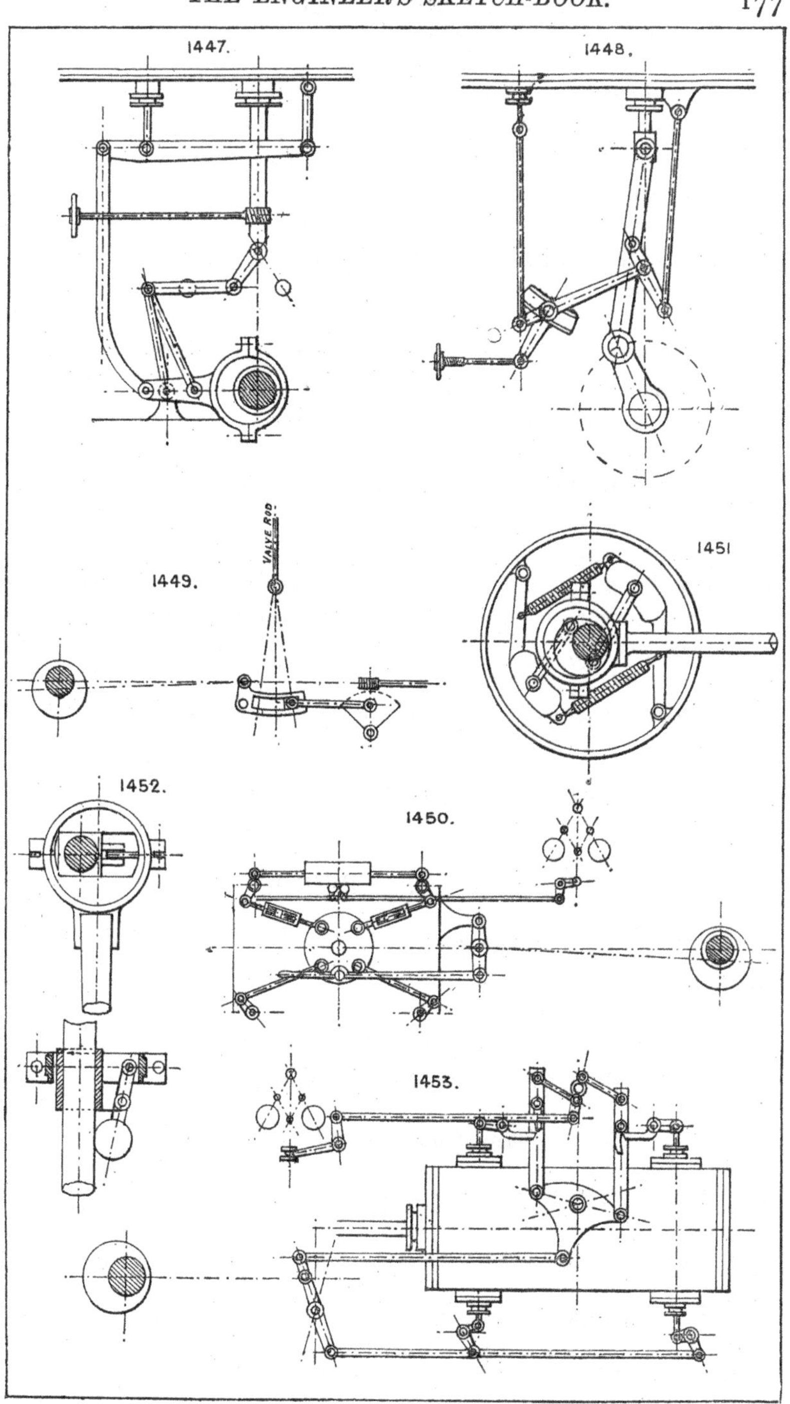
1447.
1448.
VALVE ROD
1449.
1451
1452.
1450.
1453.

1454. **Double angular slide valve** for varying the cut off by a transverse motion given to the valve from outside, either by hand gear or by the governor, the valve being made wider than the valve face, as dotted lines.

1455. **H. Jack's variable expansion gear,** with one eccentric. Patent No. 4167/85.

1456. **Variable cut-off valve** on the back of the main slide, the rod of which can be revolved by hand or from the governor to vary the opening of the cut off valves.

1457. **A plan to effect the same object** but by a cylindrical cut off valve.

1458. **English's expansion gear.** Two eccentrics. The expansion valve has no lap, and the gear gives a constant relative motion to both valves.

1459. **Tappet gear,** sometimes used for water-pressure engines, &c.

Section 80.—SPRINGS.

1460. **Open spiral spring** for tension.

1461. **Close spiral spring.**

1462. **Open spiral spring** for compression.

1463. **Open spiral spring** (square thread) for compression.

1464. **Double volute chair spring.**

1465. **Spindle-shaped open or close spiral spring** for tension.

1466. **Parallel open or close spiral spring** with coned ends.

1467 & 1468. **Volute springs.**

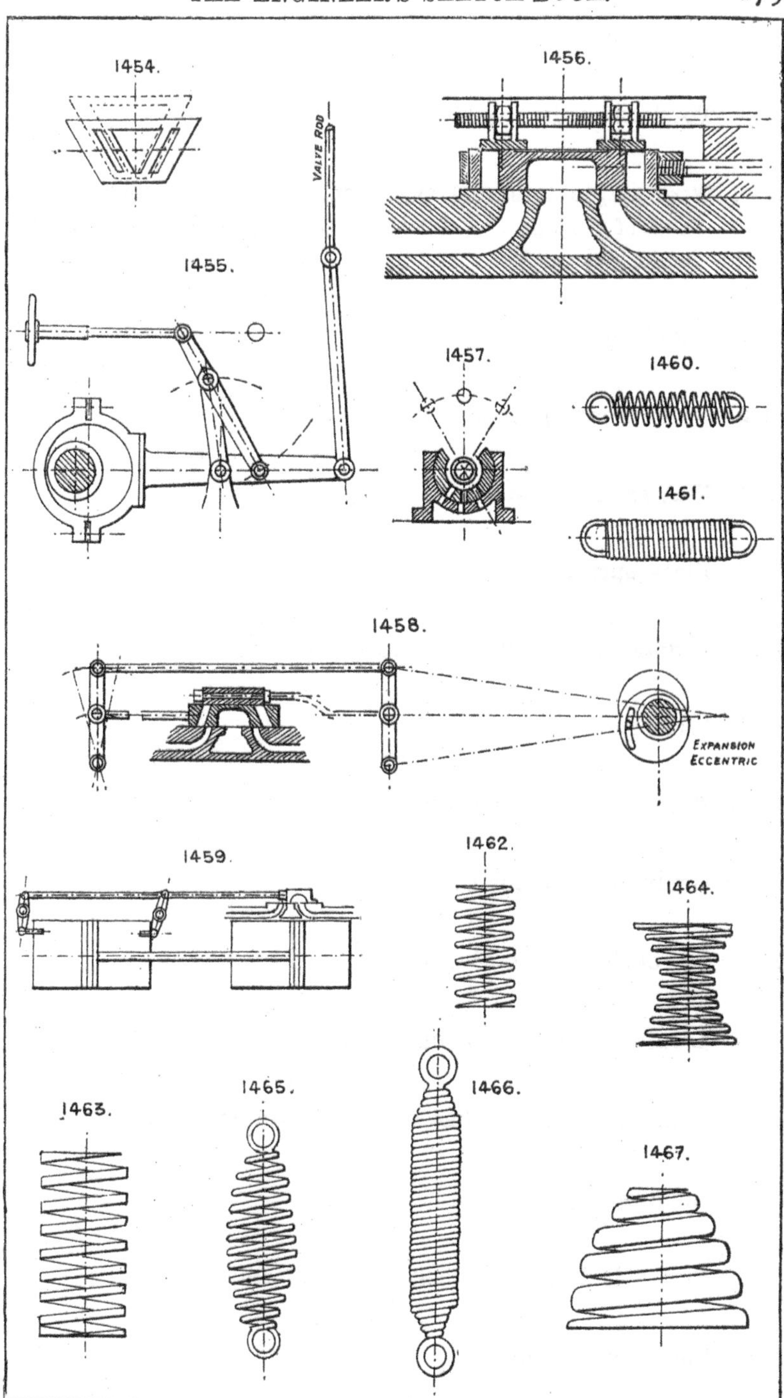
1454.
1456.
VALVE ROD
1455.
1457.
1460.
1461.
1458.
EXPANSION
ECCENTRIC
1459.
1462.
1464.
1465.
1466.
1463.
1467.

1469. **Torsional spiral spring.**

1470. **Wire staple torsional spiral spring,** used for hinging; the ends of the wire are bent at a right angle and driven into the wood.

1471. **Fixed spring.**

1472 & 1473. **Sear springs.**

1474. **Flat spiral spring.**

1475. **Plate spring.**

1476. **Indiarubber spring** for tension.

1477. **Wire spring.**

1478. **Ribbon spring** for torsion.

1479. **Compound rubber disc spring.**

1480. **Air cushion or spring piston.**

1481. **Laminated plate wagon spring.**

1482. **Compound dished disc or bent plate spring.**

1483. **Loop spring.**

1484. **Flat spiral spring, clock spring or coil spring.**

1485. **Split ring spring.**

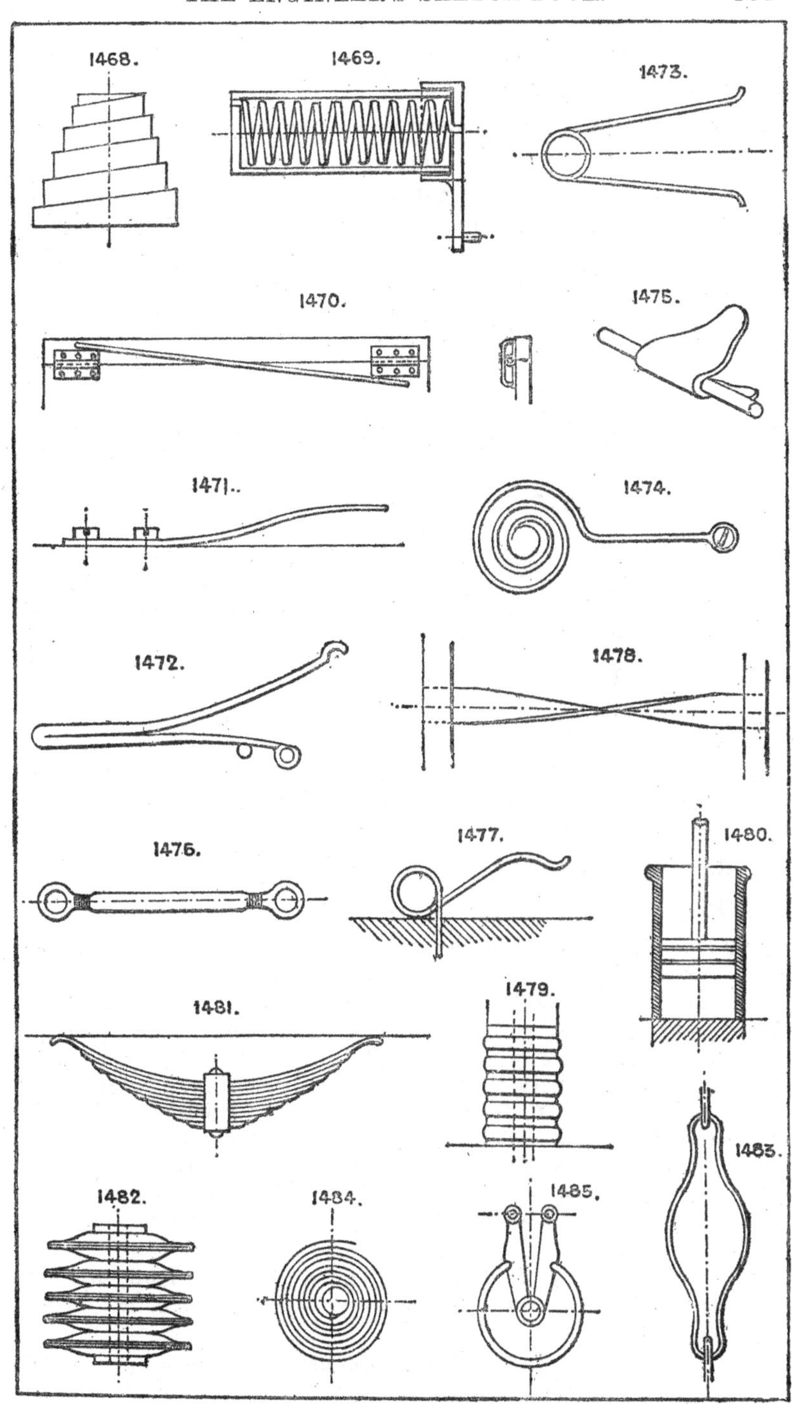
1468.
1469.
1473.
1470.
1475.
1471..
1474.
1472.
1478.
1476.
1477.
1480.
1481.
1479.
1482.
1484.
1485.
1483.

1486. **Spring pole** for foot hammer motion, &c.

1487 to 1489. **Spring washers.**

1490. **Spindle-shaped compression spring.**

1491. **Flat spiral spring** for piston rings.

See also Nos. 1729, 630, 1501, 1503, 11, 767, 768; and Section 35 (Elastic Wheels) for other forms of springs and applications of them.

For equalising the tension of a spring, see No. 1592 and 1602.

Section 81.—SAFETY APPLIANCES FOR VARIOUS USES.

FOR HOIST CAGES, &c.

1492. **Cam gear**; operates by gripping the wood guides by a serrated eccentric cam surface on the breakage of the rope, the cam being pulled round by a spring which is kept out of action by the tension of the rope until it breaks.

1493. **Strut or pawl gear**; explains itself.

1494. **Double wedge gear.**

1495. **Governor gear.** The rope attached to the cage drives a governor acting on a brake or catch which is thrown into action if the cage gains excessive speed, used in Attwood Beaver's Patent; American Elevator Co., &c.

1496. **Rack and pawl gear.**

1497. **Cross grip lever gear.**

1498. **Safety hook** to prevent accident from overwinding; the projecting horns A A strike the edges of the plate B, and throw the shackle C at top out of gear.

For hoist doors the best appliance is an ordinary spring lock opened only by a key, the doors being provided with springs to close them. Various automatic doors, revolving shutters, and other devices have also been tried. A simple and effectual protection is a continuous open-work screen wound upon a roller at top and bottom of lift, and attached to the top and bottom of cage and rising and falling with it, so that the doors into lift are all covered at all times except the one at which the cage happens to stand.

Safety valves (see Section 89). Various automatic alarm signals are applied to boilers to warn against low water or excessive pressure.

Automatic valves and other devices are applied to pumping and steam engines to prevent running away. See note to Section 41.

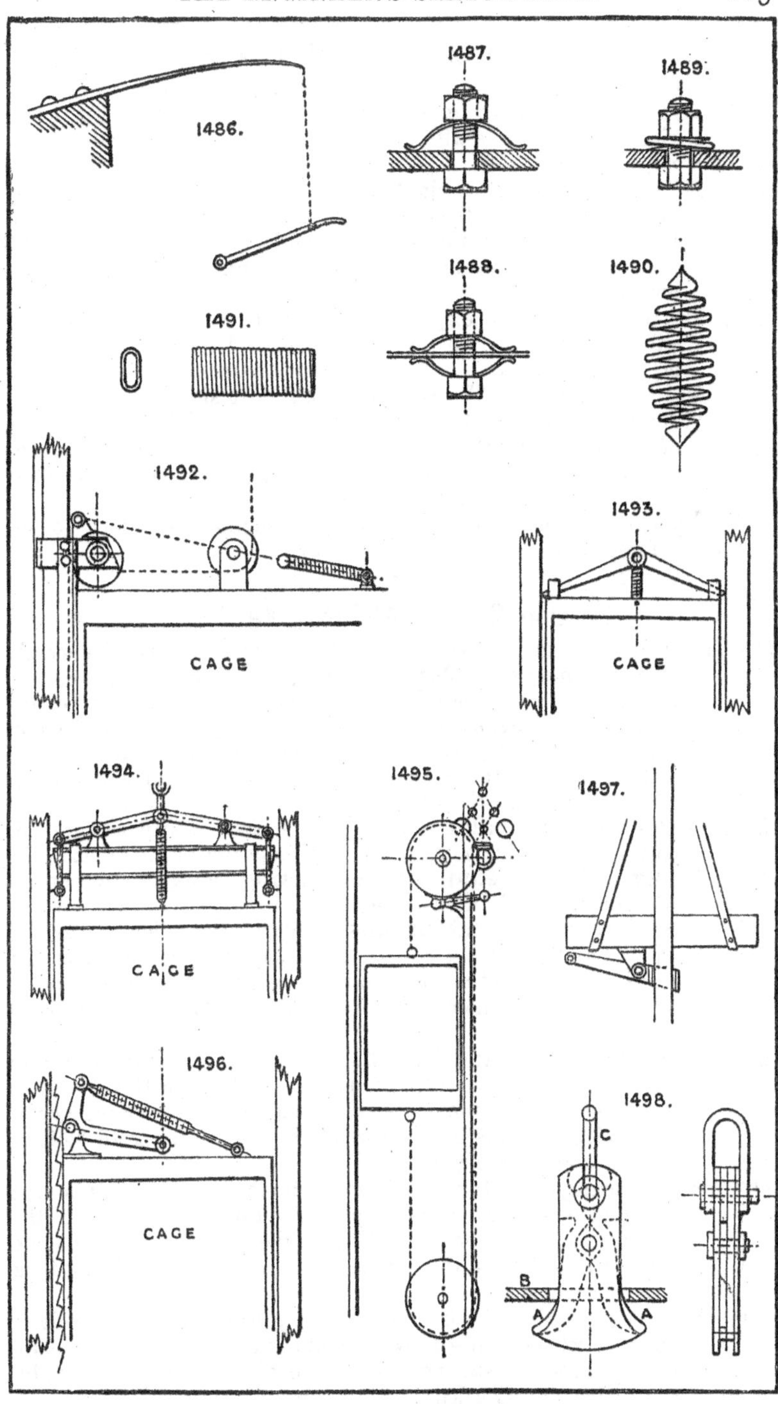
1486.
1487.
1489.
1488.
1490.
1491.
1492.
CAGE
1493.
CAGE
1494.
CAGE
1495.
1497.
1496.
CAGE
1498.
C
B
A
A

Section 82.—STEAM TRAPS.

To collect and discharge condensed steam from pipes, &c.

1499. **Trap operated by a modified form of ball cock,** which rises as the box fills with condensed water and opens the discharge valve.

1500. **Effects the same object by a floating basin.** The condensed water enters the box outside the basin, fills it, and lifts the basin which closes the discharge outlet; when the box is full the water overflows into the basin and sinks it, thus opening the outlet valve.

1501. **Trap operated by expansion** of a bent spring which closes the valve, on the principle that live steam is hotter than the water condensed from it.

1502. **Tredgold's trap.** The valve is opened by a simple float.

1503. **Wilson's trap,** like 1501, is dependent on the different expansion of a spring under the difference of temperature of the steam and condensed water. In this case the spring is formed of a steel and brass plate riveted together.

There are many other forms upon similar principles to the foregoing.

Section 83.—STARTING VALVES.

The valves used for starting steam and other engines are usually merely of the ordinary screw down or sliding types. See Section 89.

For starting and controlling all other forms of reciprocating cylinder motors, such as hydraulic lifting cylinders and presses for all purposes, the ordinary slide valve with either two or three ports is the common device; also the ordinary three or four way cock. See Section 89.

1504. **Locke's 3-way balanced valve,** which is balanced in all positions. A is the supply, B the cylinder branch, and C exhaust.

1505. **Fenby's 3-way equilibrium starting valve.** A supply, B cylinder branch, C exhaust.

1506. **Auxiliary valve and pistons** to start large slide valves too heavy for direct hand power. A 3-way cock is shown as the auxiliary valve, but a small slide valve or piston valve may be substituted. See note at foot of Section 93, also Nos. 1740 and 1741.

1507. **Auxiliary valve and bellows** for air, as sometimes used in large organs to open heavy "pallets." The small valve A is opened by the pressure of the finger on the corresponding key of keyboard, and allows the pressure of air to enter the small bellows which operates the large valve B.

1508. **Four plunger valve,** used for double power hydraulic lift cylinders employing a trunk piston. For the low power the pressure water acts on both sides of the piston; for the double power it acts only on the back of piston, the front side being then open to the exhaust.

1509. **A starting valve,** having two ordinary wing or spindle valves, either of which is lifted by a double cam or wiper on a spindle passing through a stuffing box on side of valve case. A supply, B cylinder port, C exhaust.

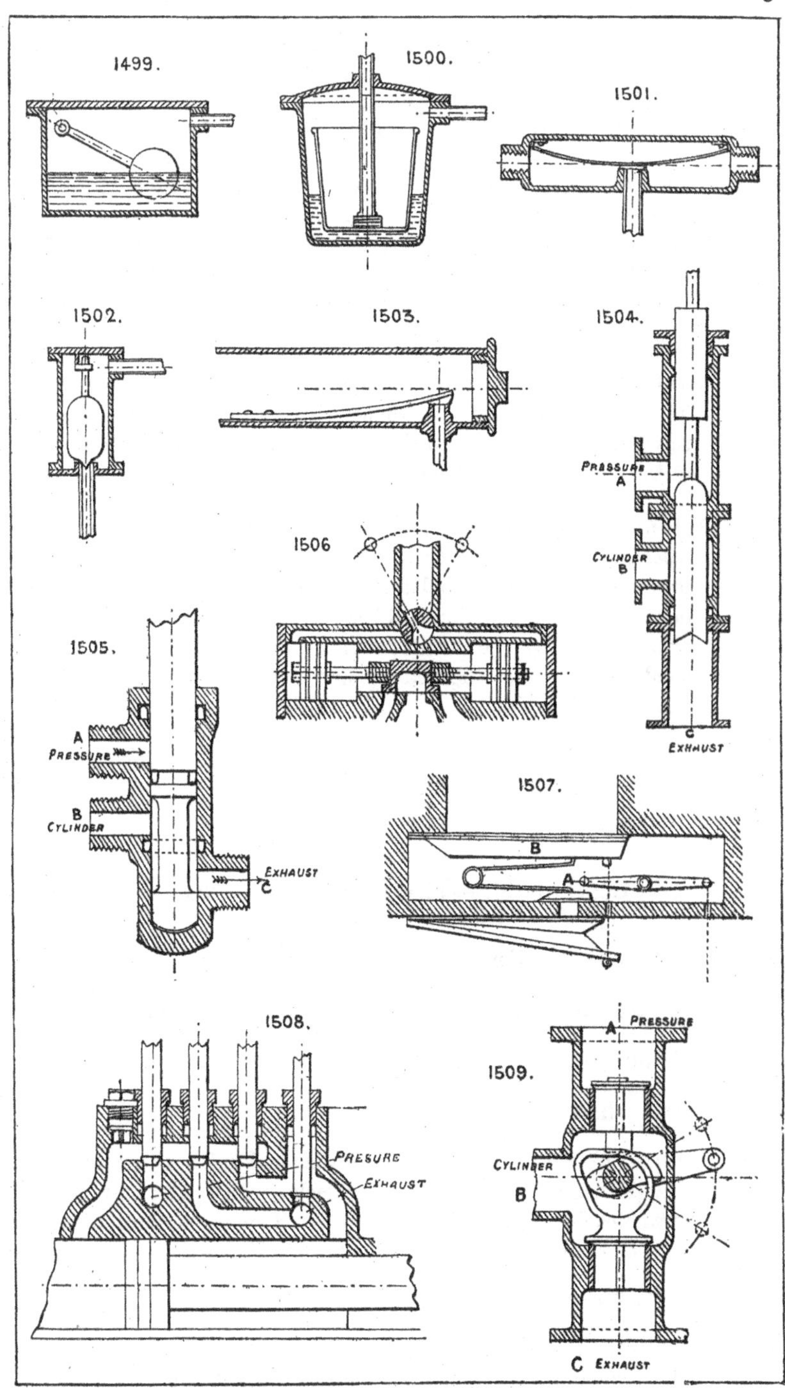
1499.
1500.
1501.
1502.
1503.
1504.
PRESSURE
A
CYLINDER
B
C
EXHAUST
1506
1505.
A
PRESSURE
B
CYLINDER
EXHAUST
C
1507.
B
A
1508.
PRESURE
EXHAUST
1509.
A PRESSURE
CYLINDER
B
C EXHAUST

1510 & 1511. **Two methods of operating starting valves** for hydraulic lifting machines. 1510 acts by counter balance weights and a single rope, each of the weights being heavy enough to move the valve, and 1511 by an endless rope.

1512. **Low pressure starting valve,** used for piston hydraulic cylinders in which the lifting is performed by the down stroke of the piston rod, and in lowering, the valve allows the water to pass from above to below the piston, the water being exhausted from below the piston on its down stroke when the valve is in the position shown.

1513. **Oscillating valve,** with plunger face kept up by a spring. The oscillating valve has two ports passing out at opposite ends, through stuffing boxes, to either end of the cylinder; the inlet is at top and discharge at bottom.

1514. **Balanced self-acting starting valve,** suitable for large machines and low pressure. The upper piston is larger than the lower or main piston, and the space above the upper piston can be put in communication with the pressure water below it, or the exhaust port by the small piston valve at top operated by the hand gear; so that the main piston is operated by the pressure water acting on it.

Section 84.—TOOTHED GEARING.

1515. **Spur gearing.** For construction of teeth see text books.

1516. **Strongest form of spur teeth** for motion in one direction only.

1517. **Half shrouded spur teeth.**

1518. **Whole shrouded spur teeth.**

1519. **Double helical spur teeth,** stronger by 15 per cent. than straight teeth; work without backlash or noise, and may be half or wholes hrouded; section of tooth on plane of motion is the same as the ordinary spur teeth (No. 1515).

1520. **Crown wheel and pinion.**

1521. **Long teeth spur wheels or "star" wheels.** Used on roller mangles, &c., where the centres rise and fall.

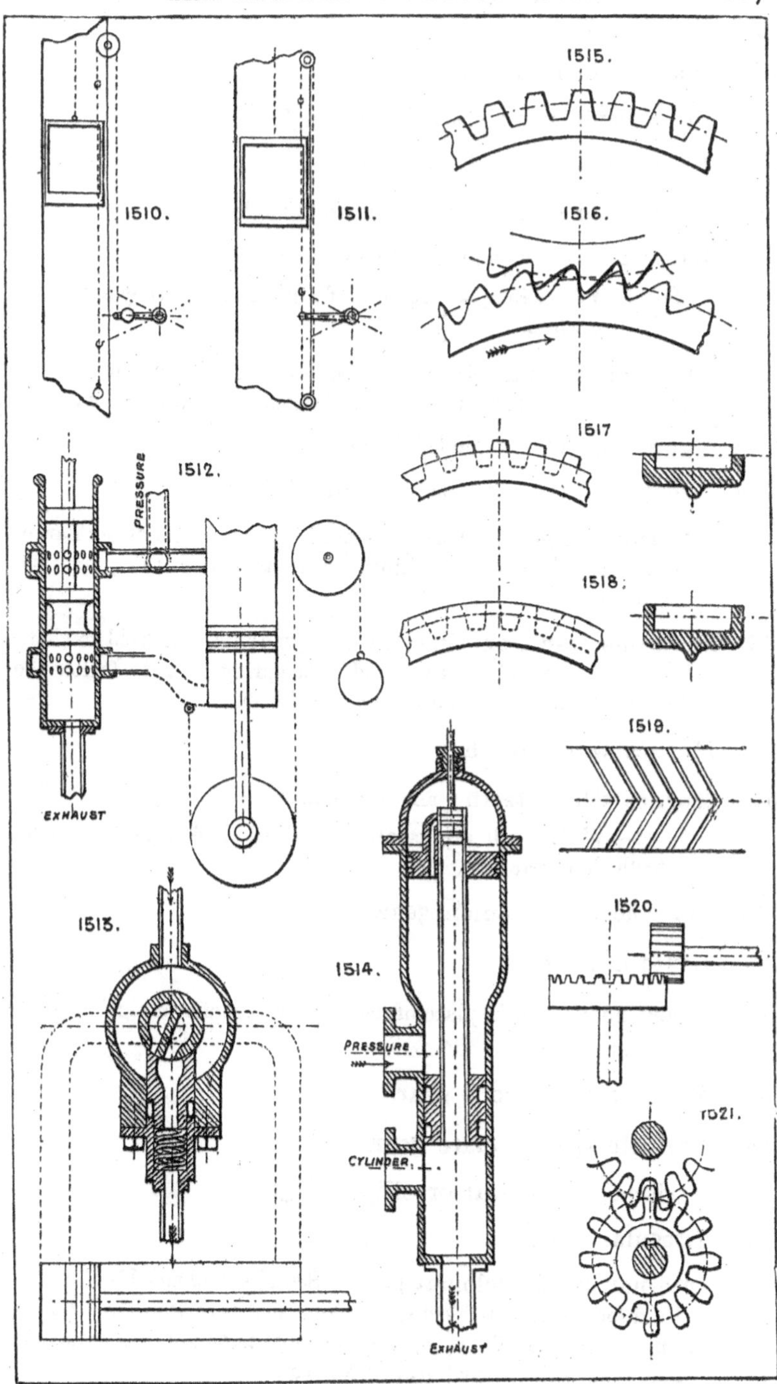
1510.
1511.
1515.
1516.
1517
1518.
1512.
PRESSURE
EXHAUST
1519.
1520.
1513.
1514.
PRESSURE
CYLINDER.
EXHAUST
1521.

1522. **Plain bevil gear;** shafts at right angles.

1523. **Plain bevil gear;** shafts at acute angles.

1524. **Plain bevil gear;** shafts at obtuse angle.

1525. **Plain bevil gear;** four shafts at right angles.

1526. **Skew bevils;** shafts not in line with one another.

NOTE.—Where the pair are both of same diameter they are called "mitre wheels."

1527. **Spur wheel and pinion;** to increase or decrease power and speed the diameters can be varied to almost any proportion.

1528. **"Screw gear";** single helical gear.

1529. **Skew spur wheels;** shafts not parallel.

1530. **Dr. Hooke's gear.** Three or more separate wheels of similar or dissimilar pitch fixed together so as to divide the pitch and reduce backlash.

1531. **The same result obtained by two wheels,** one fixed to shaft, the other loose and forced round by a spring so as to follow the pitch of the pinion and destroy all backlash.

1532. **Mortise wheel teeth.**

1533. **Mortise wheel teeth;** another method.

NOTE.—Wood teeth are usually one-third thicker than the iron teeth they gear into.

1534. **Pin wheel and pinion gear.**

1535. **Lantern wheel.**

1536. **Screw gear,** used in place of bevil gear. Shafts at right angles; teeth at an angle of 45°.

1537. **Variable speed cone gear.**

1538. **Variable speed square gear.**

1539. **Variable speed oval or elliptical gear.**

1540. **Irregular gear.**

1541. **Internal or epicycloidal gear.** See Nos. 550 and 1545. Used for differential blocks, &c. Note that both wheel and pinion run in the same direction, and that more teeth are in gear at one time than with external gear as No. 1527.

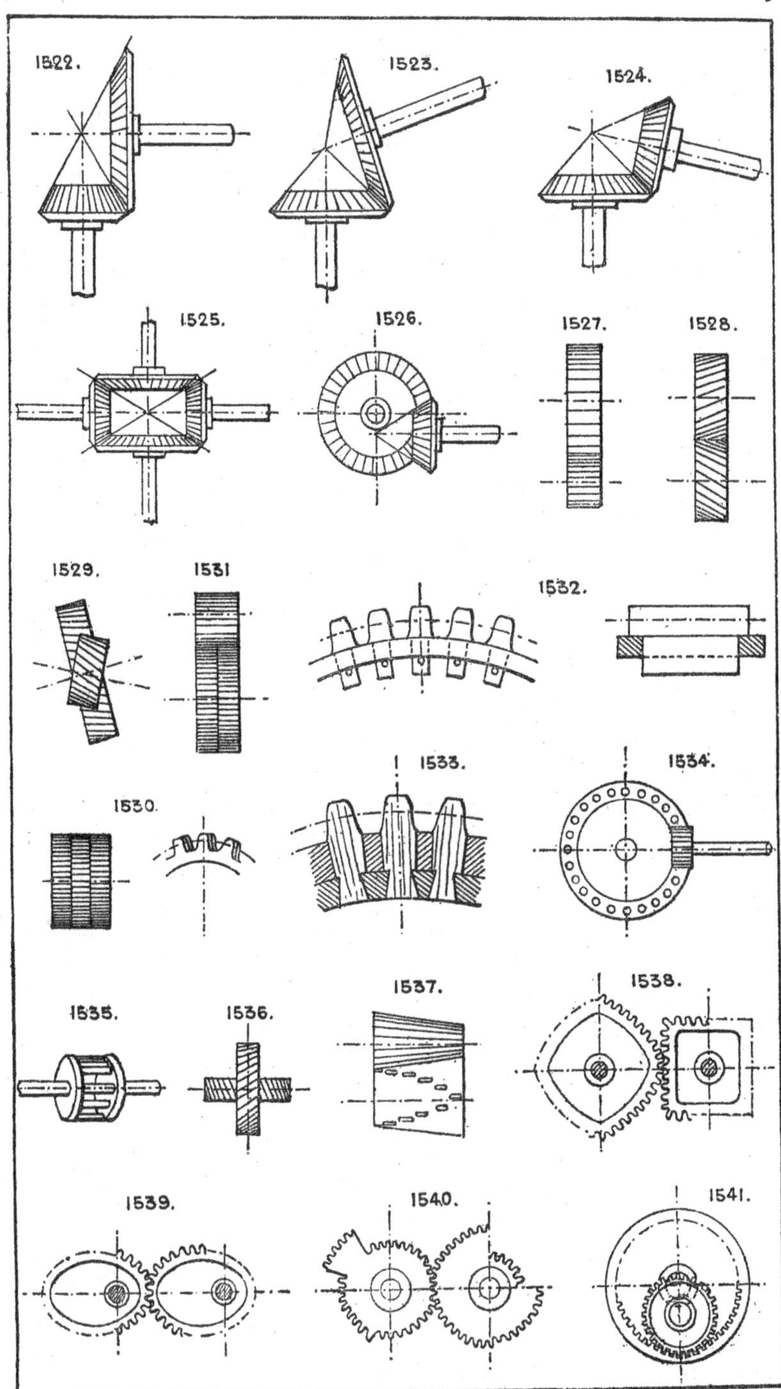
1522.
1523.
1524.
1525.
1526.
1527.
1528.
1529.
1531
1532.
1530.
1533.
1534.
1535.
1536.
1537.
1538.
1539.
1540.
1541.

1542 & 1543. **Varieties of "mangle" gear.** The pinion being revolved continuously in one direction produces a reciprocating motion of the wheel; the pinion shaft travels from inside the wheel to outside, and *vice versâ*, by rising and falling in the slot in the frame. See also No. 423.

1544. **Differential gear.** See Section 31. One wheel has one or more teeth more than the other; used for counters, &c.

1545. **Moore's patent differential epicycloidal gear.** The pinion and wheel are loose on the shaft and eccentric. One wheel has one tooth more than the other.

1546. **Multiplying bevil gear.** A is a fixed wheel, the cross C is keyed to shaft, B loose on ditto, D and E loose on C; then B is driven at a speed greater than the shaft in proportion to the diameters of the gear. See Patent No. 12,696, 1884.

1547. **Double worm gear,** right and left hand threads. Neutralises the end thrust on shaft. A and B may be geared together.

1548. **Pointed gear**; used for light work and for minimum of friction.

1549. **Curved worm gear,** for heavy strains. Several teeth are in gear at once, but the thread, having a varying section and pitch, is difficult to cut.

1550. **Antifriction worm gear** (Hawkins'). The wheel has four rollers; when one pair is nearly out of gear with the worm, the next pair is coming into gear. This worm is also difficult to cut.

1551. **Crown worm gear.**

1552. **Ball joint mitre gear.**

1553. **Multiplying rack gear.** The upper moving rack is driven at twice the speed of the spur wheel rod. The lower rack is fixed; used on planing and printing machines.

1554 to 1557. **Varieties of worm gear,** with straight, hollowed, and curved teeth; the latter are strongest.

1558. **Worm and rack gear.**

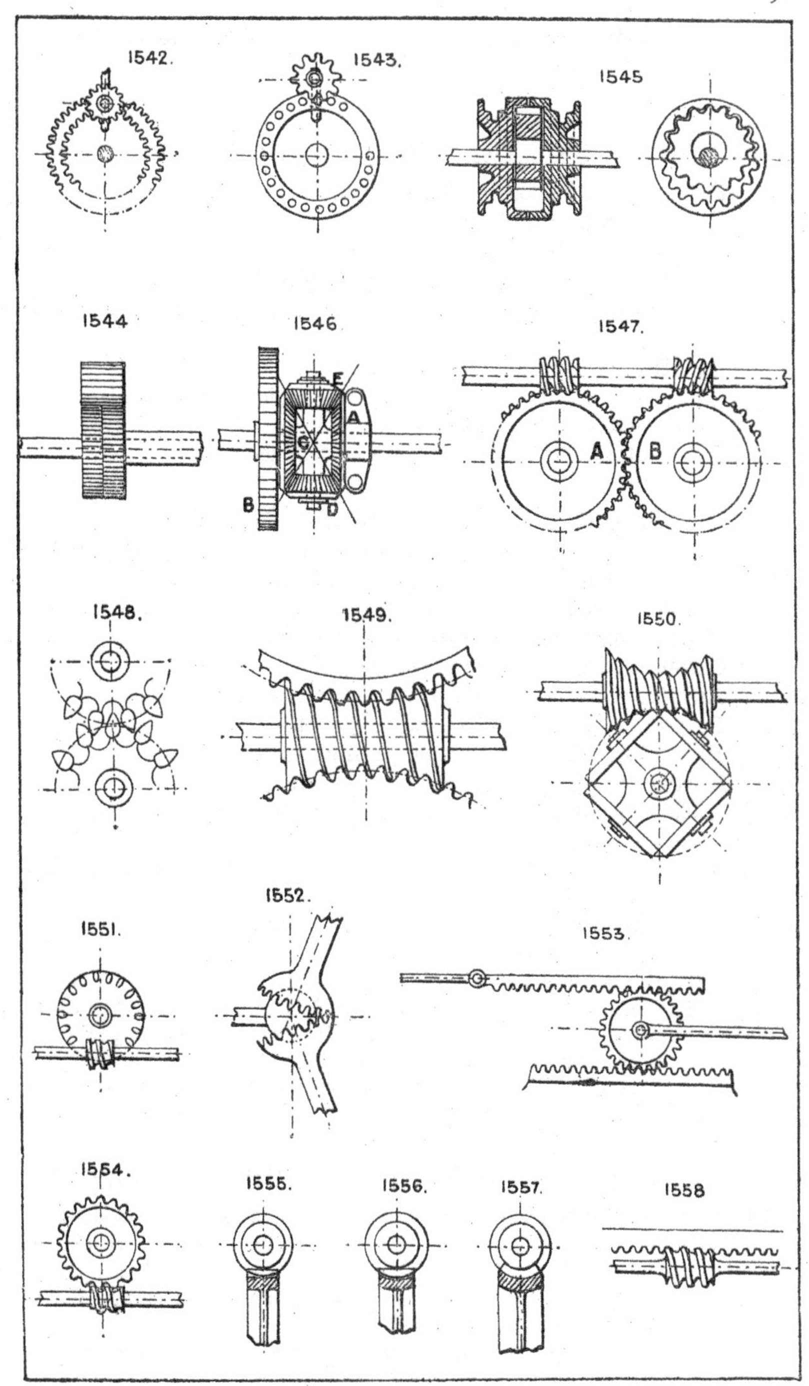
1542.
1543.
1545
1544
1546
E
A
C
B
D
1547.
A
B
1548.
1549.
1550.
1551.
1552.
1553.
1554.
1555.
1556.
1557.
1558

1559. **Differential worm gear.** The worm gears into two wheels, one having one tooth more than the other.
See also Sections 40 and 31.

Section 85.—TRANSMISSION OF POWER.

a. **By belt, chain, or rope.** See Sections 3, 66.
b. **By shafting.** See Section 76.
c. **By gearing.** See Sections 84 and 40.
d. **By steam or air** conveyed in pipes (elastic fluids).
e. **By water, glycerine, or oil** conveyed in pipes (non-elastic fluids).
f. **By stiff rods** running over guides.
g. **By wires or ropes** running over guide pulleys—wire rope transmission. See Section 66.
h. **By electricity** conveyed along wire conductors.

Section 86.—TANKS AND CISTERNS.

1560. **Plan of square tank** of ordinary form; formed of cast iron flanged plates and wrought iron tie rods, the joints are either made with rust cement or planed and jointed with tape and red lead.
1561. **Plan of square tank** with rounded angles.
1562. **Circular tank.** No tie rods required.
1563. **Elliptical tank.** Requires tie rods across the flat sides.
1564. **Polygonal tank.** No tie rods required.
1565. **Elevation of square or polygonal tank.**
1566. **Elevation of cylindrical or circular tank.**
1567 & 1568. **Condensing or cooling tanks.** Surface condensers with sloping trays or tubes.
1549. **Wrought iron tank,** usual section, formed of sheets and L irons riveted together.
1570 & 1571. **Circulating or depositing tanks.**
1572. **Boiler saddle tank.**
1573. **Circulating tank** for hot water.
The level of water may be maintained in tanks by either an overflow pipe or notch, or by a ball cock on the supply pipe. Glass water gauges can be fixed outside to show the level of the water inside; and floats are used, attached to a cord and pulleys, for the same purpose. See also No. 1730.

Section 87.—THROWING IN AND OUT OF GEAR.

1574. **The driving wheel** is loose on shaft, and is locked to its shaft by the hand wheel nut (see No. 945), or by a ratchet wheel and locking pawl.
1575. **Two half-nuts** are lifted in or out of gear with the screw by cam or lever motion. See No. 942.
1576. **One shaft runs in eccentric bearings,** which can be revolved so as to throw it out of gear with the other shaft.

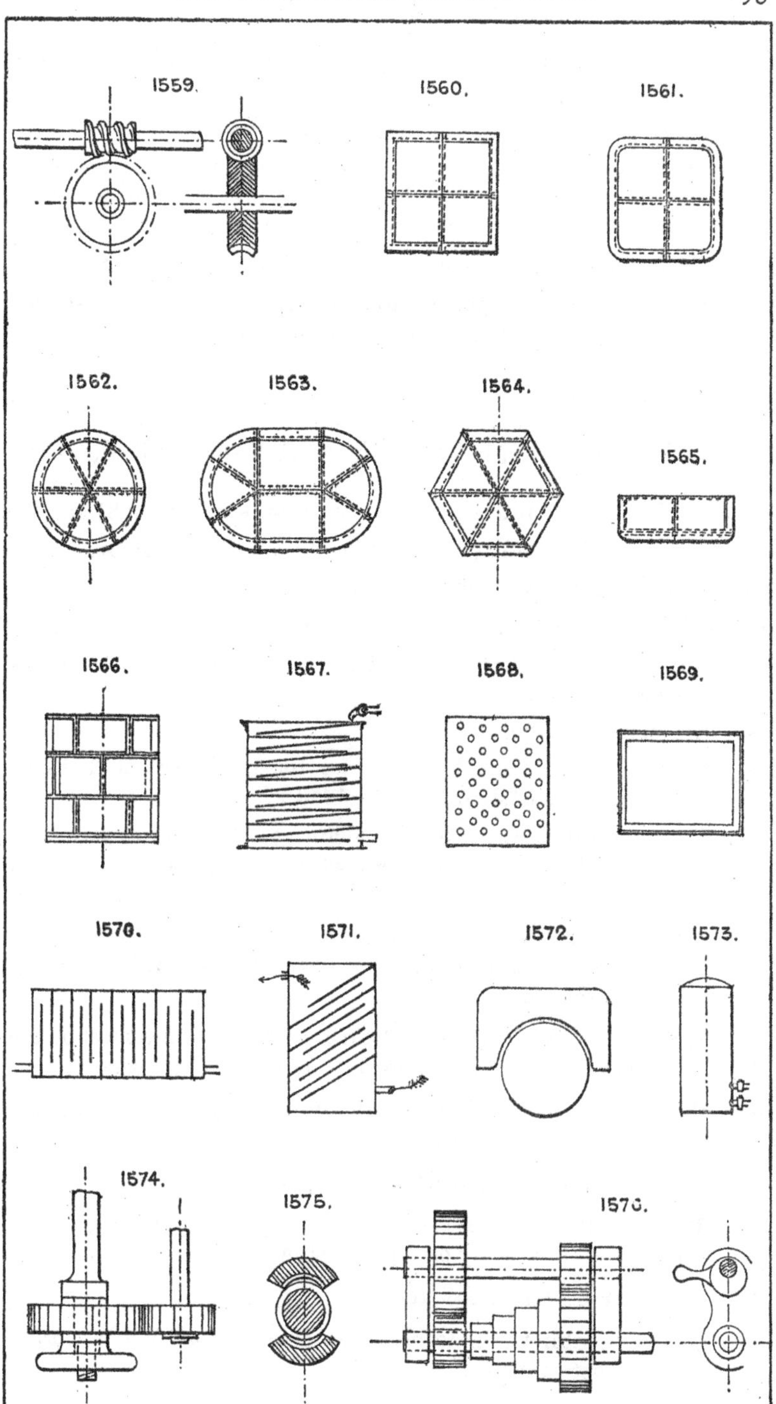
1559.
1560.
1561.
1562.
1563.
1564.
1565.
1566.
1567.
1568.
1569.
1570.
1571.
1572.
1573.
1574.
1575.
1576.

1577. **Radius bar and slot.** The stud wheel can be shifted in or out of gear along the slot.

1578. **Sliding back shaft,** slides out of gear. See dotted lines.

1579. **Method of throwing a pulley out of gear** by slacking the belt. This is done either by a cam bearing or sliding motion to the driven shaft. Works best vertically. See also No. 1219.

1580. **Cam slot motion** for back shaft. To throw it in or out of gear.

1581. **Motion employed for punching machines,** &c. To set the punch in or out of action by a cam and hand lever.

Section 88.—VARIABLE MOTION AND VARIABLE POWER.

For variable speed and power by spur gear, see Sections 84 and 40.

For variable speed and power by bevil gear, see Section 84.

For variable speed and power by cam gear, see Section 9.

For variable speed and power by belt gear, see Section 3.

1582. **Variable speed belt cones,** for crossed belts. Angle of cones should not exceed 15°.

1583. **Stepped cone gear.**

1584. **Variable throw crank pin.** (See Hastie's Patent, 3561, 1878; Knowelden and Edwards', 2996, 1858.)

1585. **Beam motion,** with variable fulcrum to alter the proportionate lengths of stroke of driving and driven cylinders. See also No. 1606.

1586. **Variable pressure accumulator.** Both cylinders are in communication by a pipe, and the pressure varies with the angle between the rams.

1587. **Wright's variable gear.** The radius of frictional contact of the wheels varies as they are moved closer together or separated.

1588. **Olmsted's variable cone friction gear,** with intermediate double cone idle wheel instead of belt.

1589. **Convex and concave cones** for *open* belts.

1590. **Three speed gear,** each separate pair of spur gear being driven by its own belt pulley on separate sleeve pieces.

1591. **Irregular or elliptical gear.**

1592. **Lever combination** to obtain an even tension from a spring throughout its motion.

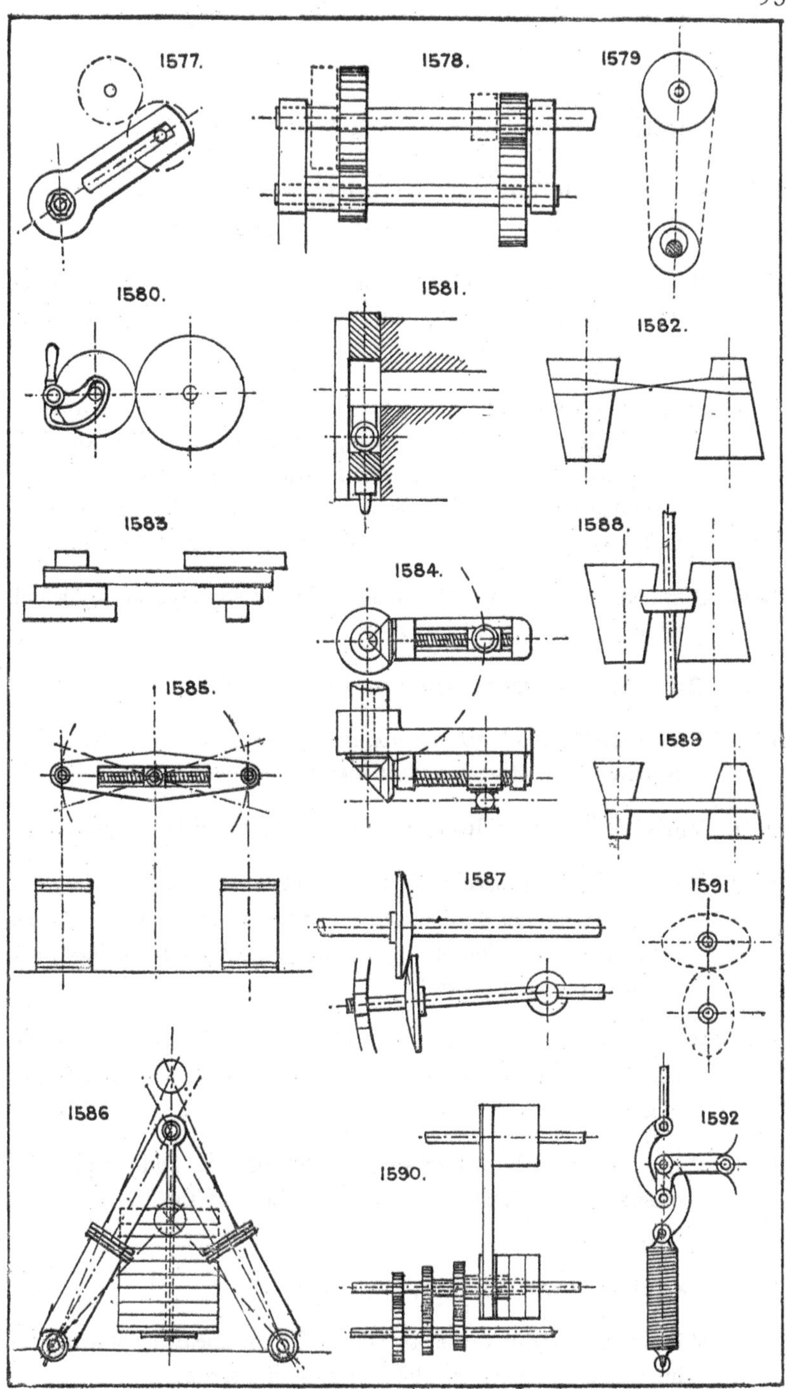
1577.
1578.
1579
1580.
1581.
1582.
1583
1584.
1585.
1586
1587
1588.
1589
1590.
1591
1592

1593. **Scroll spur gear.**

1594. **Scroll gear** for obtaining a variable pull from a weight.

1595. **Variable friction gear.** The pinion can be moved up or down from the centre to the outside of disc to vary the speed.

1596. **Owen's compound lever variable pressure air pumps.** The pressure increasing and speed decreasing as the pistons rise to the top of their stroke.

1597. **2-speed gear** by one belt. One loose pulley B carries a transverse mitre wheel, which, gearing into the fixed mitre wheel A, drives the mitre wheel pulley C keyed to shaft at twice the speed of pulley B when the belt is on B; by shifting the belt to pulley C, the speed is 1 to 1. The third pulley is loose for running idle.

1598. **2-speed bevel gear** with three wheels and sliding shaft, by which either pair can be put into gear.

1599. **2-speed bevel gear** with four wheels and sliding shaft.

1600. **Increasing speed cone and screw,** friction gear. The cone is driven by frictional contact with the pinion.

1601. **Variable fulcrum lever,** with shifting pin and hole adjustment.

1602. **Fusee barrel,** as used in clocks and watches to equalise the tension of spring on the movement; may also be employed to give a variable speed. The spring is usually similar to No. 1484, and coiled in the upper barrel.

1603. **Variable throw crank pin and slot,** much used for variable feed motions in combination with some type of pawl and ratchet gear.

1604. **Variable travel imparted to the piston rod from the crank** by altering the point of attachment of the link A to the slot.

1605. **Variable power pistons,** single-acting.

1606. **Effects the same ends as No. 1601 or 1585,** by shifting the fulcrum point along the slot by means of a screw.

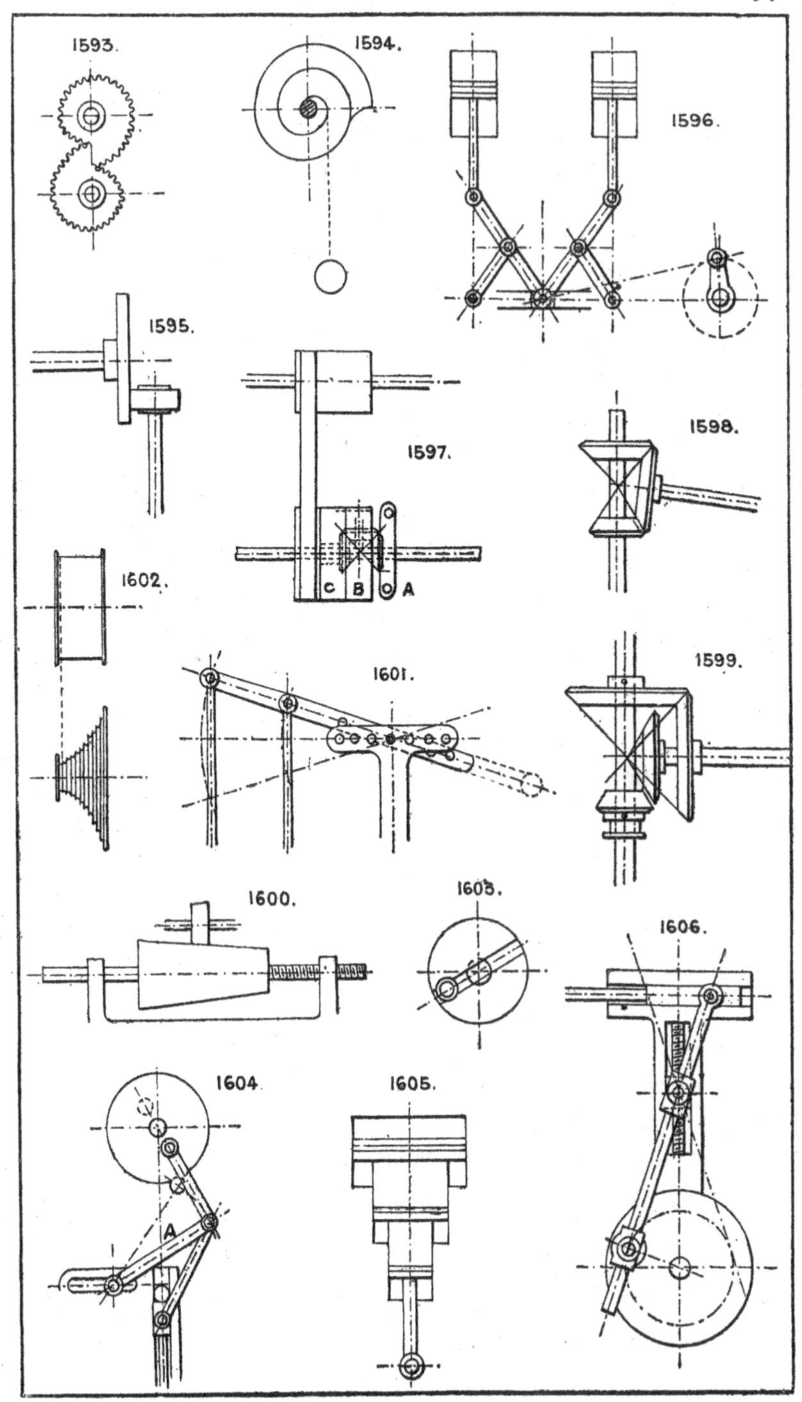
1593.
1594.
1596.
1595.
1597.
1598.
1602.
C
B
A
1601.
1599.
1600.
1603.
1606.
1604
A
1605.

1607. **Variable throw crank pin** by means of a jointed crank and radial adjusting screw.

1608. **Variable throw crank pin** by attaching the crank pin to an eccentric disc. See Section 10.

1609. **Wind motor-fan or turbine,** with variable-angle vanes actuated by a central sleeve and cam or lever gear.

1610. **Variable friction cone gear**; the small friction pinion can be moved radially to and fro to alter the leverage and consequently speed of driving radius from the cone.

In the steam engine, compressed air engine, and gas engine (these being all elastic fluids), the power given out is varied by altering the supply of steam, air, or gas. In the water wheel the power may be varied by altering the quantity of water (or head of water) supplied to the wheel. In the turbine the power can be varied by altering the head of supply water and the angle of vanes; altering the quantity reduces the speed and efficiency. The turbine will not work well under great variations of either head or quantity.

See also Nos. 736, 722, 723, 1190, 1191, 381, 382, 384, 385, 377, 372, 373; and Sections 20 and 40.

For variable pressure or tension by springs, see Section 80.

Variable balance weights, Section 20.

Section 89.—VALVES AND COCKS.

Of the almost innumerable varieties of valves and cocks in use the following are selected as types without reference to special uses, each type having its peculiar value, and the drawings are only intended to indicate the special features of each type without such details as may be varied to suit each particular requirement or application.

1611. **The common plug cock.**

1612. **The same,** but with screwed gland.

1613. **2-way or 3-way plug cock,** with packed gland.

1614. **Hollow plug blow-off cock,** with packed gland.

1615. **Back pressure or check valve,** self closing.

1616. **Ball valve and guard,** self closing.

1617. **Indiarubber disc and grating valve.**

1618. **Double flap indiarubber or leather** and grating valve.

1619. **Simple flap valve** faced with rubber or leather.

1620. **Rocking or rolling valve.** For opening and closing gradually and easily against pressure.

1621. **Roll-up valve.** For same purposes as the last named.

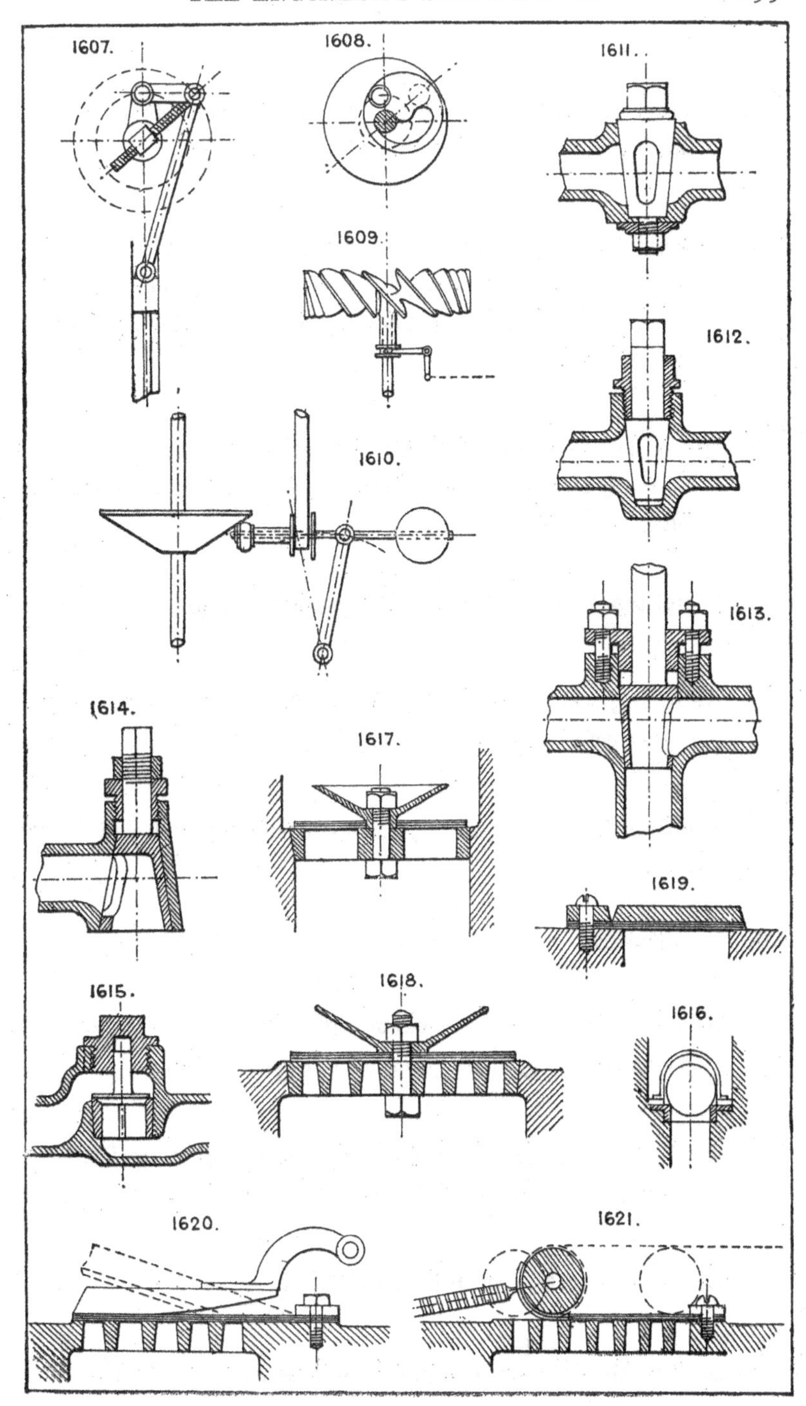
1607.
1608.
1611.
1609.
1612.
1610.
1613.
1614.
1617.
1619.
1615.
1618.
1616.
1620.
1621.

1622. **Sector full-way screw down valve,** shown closed; when open the disc is up in the chamber out of the waterway.

1623. **Double face valve;** the spindle is screwed into the stuffing-box neck and the lower valve, the upper one being pinned to the spindle; the thread in the lower valve is twice the pitch of the upper thread.

1624 & 1625. **Spring relief valves;** the springs are adjusted to blow off at any stated pressure, which may be regulated by a screw or nut (not shown).

1626. **Weighted lever relief valve or safety valve.**

1627. **Reducing valve,** can be adjusted by the balance weight to pass fluids from a high pressure to any lower pressure.

1628. **Another form,** with spring balance adjustment and equilibrium valve.

1629. **Equilibrium valve.**

1630. **Equilibrium valve,** not steam tight, but serrated, to cut off gradually, employed for governors of steam engines.

1631. **Equilibrium cylindrical grating valve,** may be used to open and close either by vertical or revolving motion.

1632. **Common throttle or butterfly valve.**

1633. **Duplex throttle or damper** for a three-way pipe or flue.

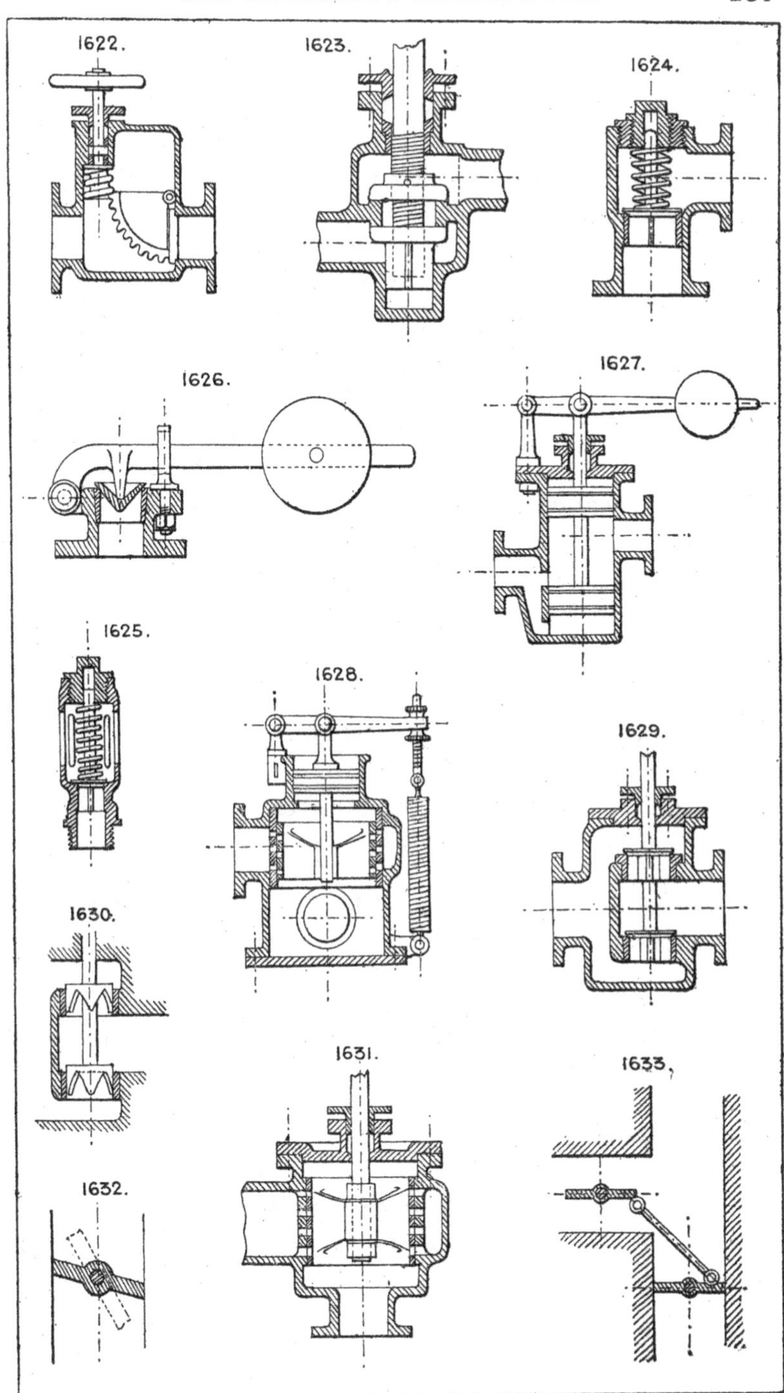
1622.
1623.
1624.
1626.
1627.
1625.
1628.
1629.
1630.
1631.
1633.
1632.

1634. **Hydraulic high-pressure check valve,** with long guide wings.

1635. **Hydraulic plug or spindle valve and seating**; all seatings for high pressure should be narrow and hard.

1636. **Duplex or Ramsbottom safety valve**; each valve serves as a fulcrum by which to lift the other. One fulcrum point should be jointed to the lever and the latter move in a vertical guide, or else the point of attachment of the spring to the lever be placed below the level of the valve seatings.

1637. **A modification** of the last named.

1638. **Pass valve,** used for pneumatic despatch tubes.

1639. **Oscillating lever duplex valve.**

1640. **Simple radial disc valve or sluice.**

1641 & 1642. **Oscillating cylindrical valves.** Corliss valves; sometimes made tapered or conical.

1643. **Multiple ball valve,** for high lift delivery valve of large pumping engines. Balls are of guttapercha, and open and close without shock.

1644. **Multiple ring valve.** The rings open and close in succession, thus avoiding shocks.

1645. **Double beat ring valve.**

1646. **Double beat equilibrium or Cornish valve.** The upper seating can be made of such area as to partly or entirely balance the valve.

1647. **Multiple ring valve.** The indiarubber rings expand and contract over the perforations.

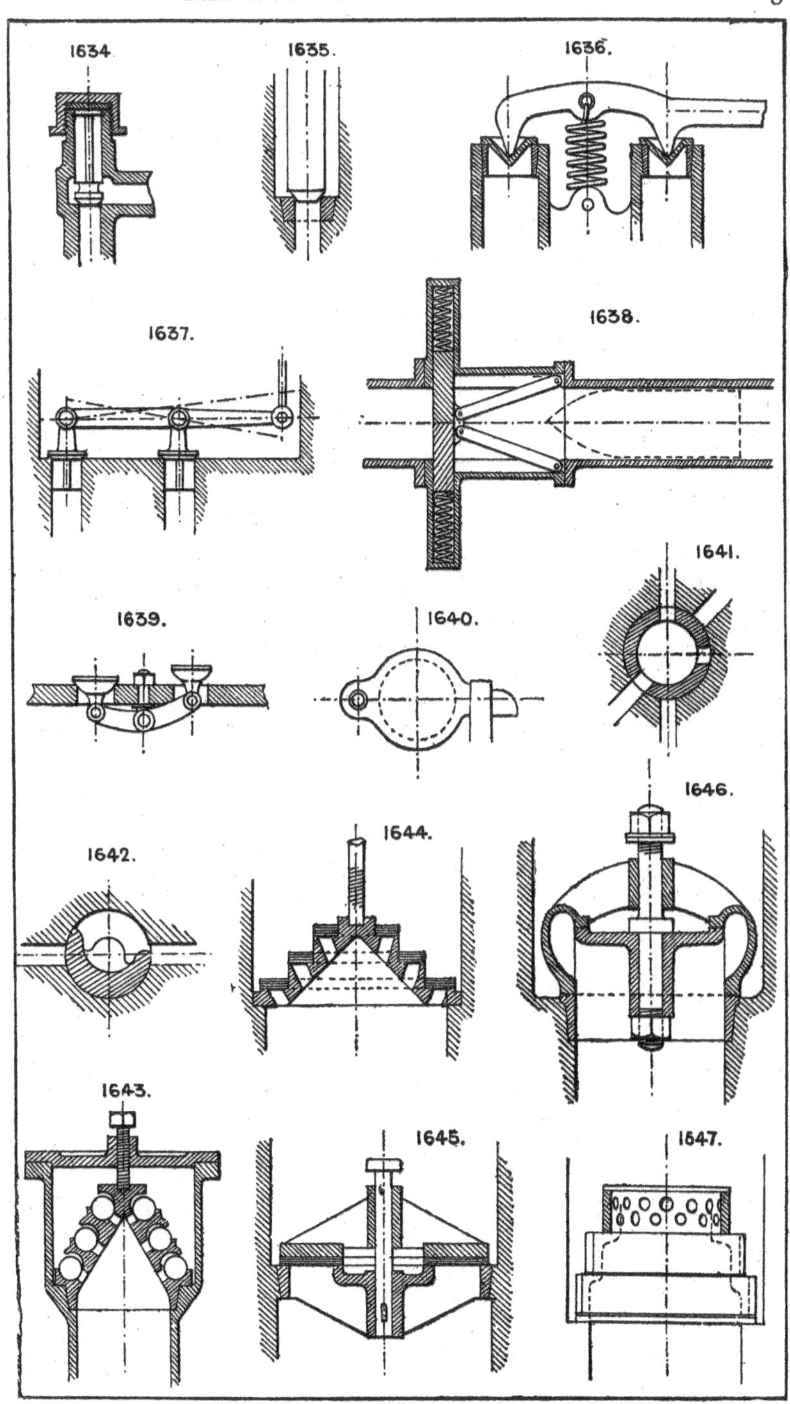
1634.
1635.
1636.
1637.
1638.
1639.
1640.
1641.
1642.
1643.
1644.
1645.
1646.
1647.

1648. **Double-beat valve,** with sunk seating.

1649. **Common D slide valve,** with three ports.

1650. **Duplex or double D slide.**

1651. **Another form,** partly in equilibrium.

1652. **Equilibrium slide valve,** with circular packed trunk on back, of area sufficient to balance the face area of valve.

1653. **Similar result** obtained by a piston and link.

1654. **Equilibrium piston valve** employed in lieu of the slide valve. A steam pipe, B B exhaust.

1655. **Gridiron slide valve.**

1656. **Common plate slide valve or sluice,** used for blast pipes.

1657. **Floating ball valve,** for automatic discharge of air from water mains.

1658. **Ordinary double-faced sluice valve,** lifts clear of the water way.

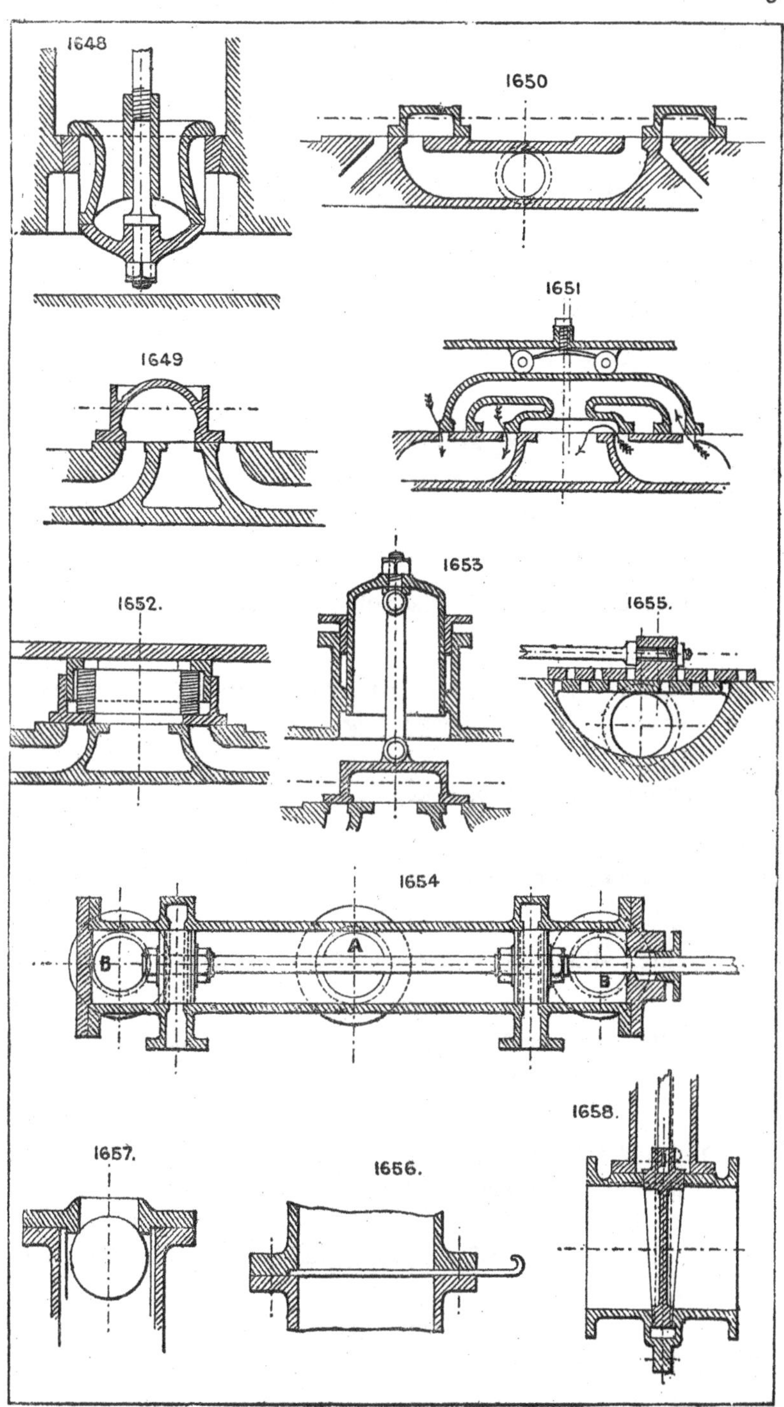
1648
1650
1651
1649
1653
1652.
1655.
1654
A
B
B
1658.
1657.
1656.

1659. **Single face sluice valve,** for sewage water, &c.

1660. **Flap sluice valve.** Tidal outlet valve.

1661. **Diaphragm valve.**

1662. **Oscillating disc valve,** for gas.

1663. **Large 3 or 4-way plug cock.**

1664. **Finger valve,** closed by a spring.

1665. **Taper cone valve,** for gradually closing an outlet.

1666. **Dome valve,** used for hot blast, hot gases, &c. This form retains its shape when heated, expanding evenly.

1667. **Common floating ball tap.**

1668. **Three-way air valve.**

1669. **Duplex slide,** used to close a number of openings at once.

1670. **West's spiral valve,** with indiarubber cord which expands and contracts over a spiral perforated groove.

1671. **Dennis' self-lifting valve.** The valve is kept to its seat by the pressure being admitted on to its back through small hole A; when the larger hole B is opened by the spindle the back pressure is relieved and the valve lifted by the pressure below its conical underside.

1672. **Common cone plug.**

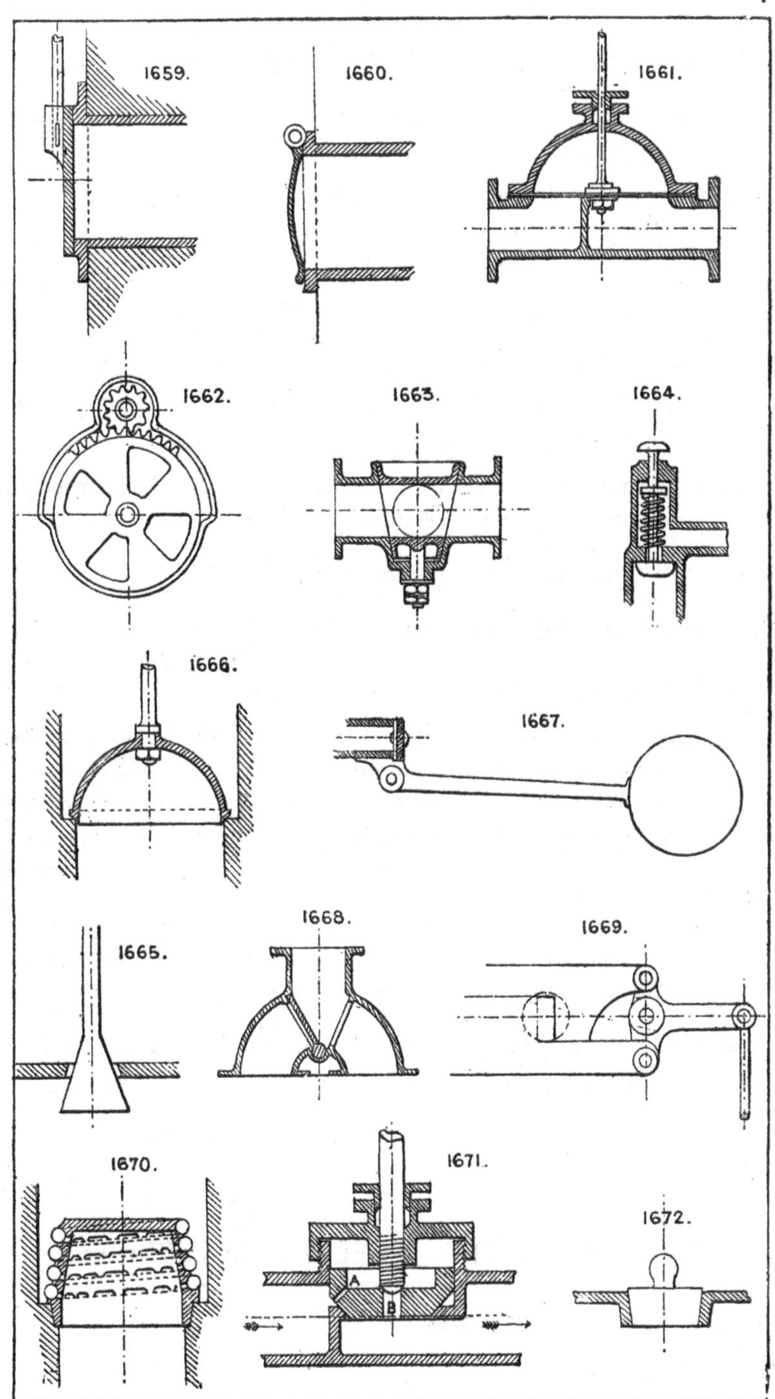
1659.
1660.
1661.
1662.
1663.
1664.
1666.
1667.
1665.
1668.
1669.
1670.
1671.
A
B
1672.

1673. **Equilibrium plug or cylindrical valve,** double ported.

1674. **Another form of self-lifting valve.** See No. 1671 for description.

1675. **Compound flap valve.**

1676. **Indiarubber pump valve.**

1677. **Venetian shutter or compound butterfly valve.**

1678. **Bye pass** used to allow a small flow when the main valve is shut off. Used also to equalise pressure on both sides of a large valve to enable it to open easily.

1679. **Bell and hopper, or cup and cone**; used for blast furnaces, coke ovens, and gas generators.

1680. **Cup valve and suspended weight.**

1681. **Four-way valve** for hot water pipes, &c.

1682. **Oscillating valve.**

1683. **Gas purifier centre valve, for four purifiers**; to work one purifier off and three on, or all four on at once. It is similar in plan to No. 1684, but has an additional top valve which allows the gas to pass into the fourth purifier; the top valve has an independent sleeve and lever motion.

1684. **Gas purifier centre valve,** employed to deliver and discharge gas into and out of any one, two, or three out of four purifiers. The motion of the gas is shown by the arrows.

1685. **Conical grating valve** with radial slots, opened or closed by revolving motion.

Section 90.—WATER WHEELS AND TURBINES.

1686. **Simple undershot wheel.**

1687. **Breast wheel.**

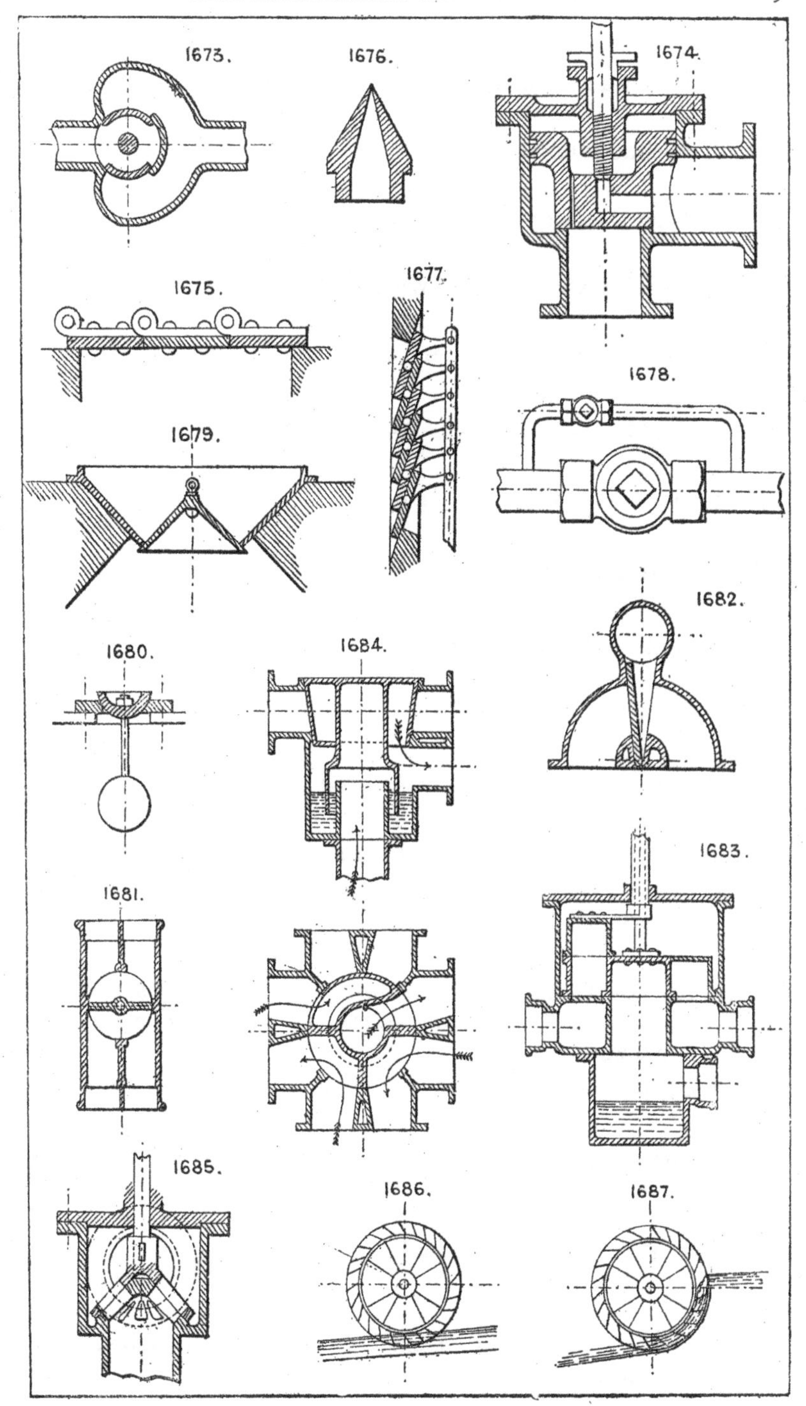
1673.
1676.
1674.
1675.
1677.
1678.
1679.
1680.
1684.
1682.
1681.
1683.
1685.
1686.
1687.

1688. **High breast wheel.**

1689. **Overshot wheel.**

1690. **Return overshot wheel.**

1691. **Internal feed re-action wheel.**

1692. **Sunk wheel,** driven by air: may be used as a meter for gas or air.

1693. **Current wheel,** driven by tidal or river current.

1694. **Flutter wheel,** with high fall.

1695. **Horizontal wheel.**

1696. **Re-action wheel,** the oldest form of turbine.

1697. **Engel's diagonal wheel.**

1698. **Scoop wheel** for raising water. See also No. 1024.

1699. **Wheel, with internal buckets** and feed.

Note that most of these may be reversed and made into water raising machines, as No. 1698.

1700 to 1703. **Sections of various forms of buckets** in wood and iron. No. 1703 is a ventilated bucket which allows air to escape as the water enters.

For governing speed of water wheels and turbines, see Section 41.

1704. **Fourneyron's turbine,** outward flow; the outer vanes are fixed, the inner ones revolve with the shaft.

1705. **Jonval's turbine,** downward flow; either the upper or lower set is fixed.

1706. **Swain's turbine,** inward and downward flow, with inward curved vanes or flumes.

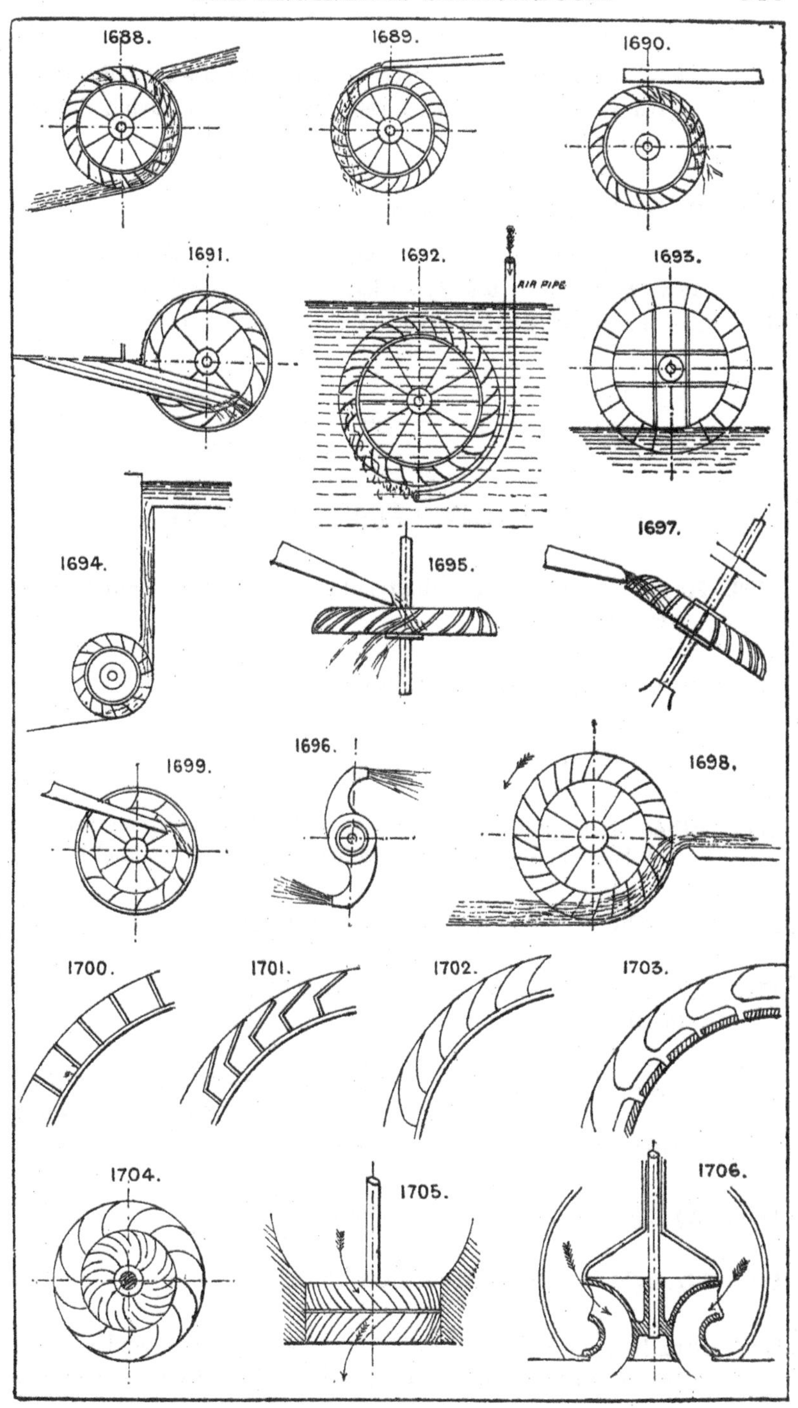
1688.
1689.
1690.
1691.
1692.
AIR PIPE
1693.
1694.
1695.
1697.
1699.
1696.
1698.
1700.
1701.
1702.
1703.
1704.
1705.
1706.

1707. **Leffel's turbine,** inward and downward flow; has one outer ring of fixed vanes and two inner sets revolving, but having different angles of flow.

1708. **Undershot jet wheel** for high pressure water.

Many other forms of turbines are extant, but are mostly modifications of the above types. The best types have means of varying the angle of the vanes and areas of passages to suit varying quantities of water.

Section 91.—WHEELS IN SEGMENTS.

1709. **Heavy gearing** for rolling mills, &c., with dovetailed joints, wedged and packed.

1710. **Wheel cast in sectors** bolted together.

1711. **Bevil wheel in halves.**

1712. **Wheel with rim in segments** bolted together, and provided with bored and cottered sockets for arms in both rim and boss.

1713. **Fly wheel rim,** cottered and dowelled together.

1714. **Arms and boss cast in one:** the rim in segments, bolted together and to the arms.

1715. **Tension wheel.** The tie spokes are sometimes arranged in two sets at a slight angle to each other to prevent the rim turning without the boss. Bicycle wheels are of this class.

1716. **Wrought-iron wheel,** with cast boss.

1717. **Wrought-iron wheel,** with cast boss.

1718. **Rim in segments** bolted together, wood arms and cast boss, with sockets to receive arms; this type is much used for water wheels.

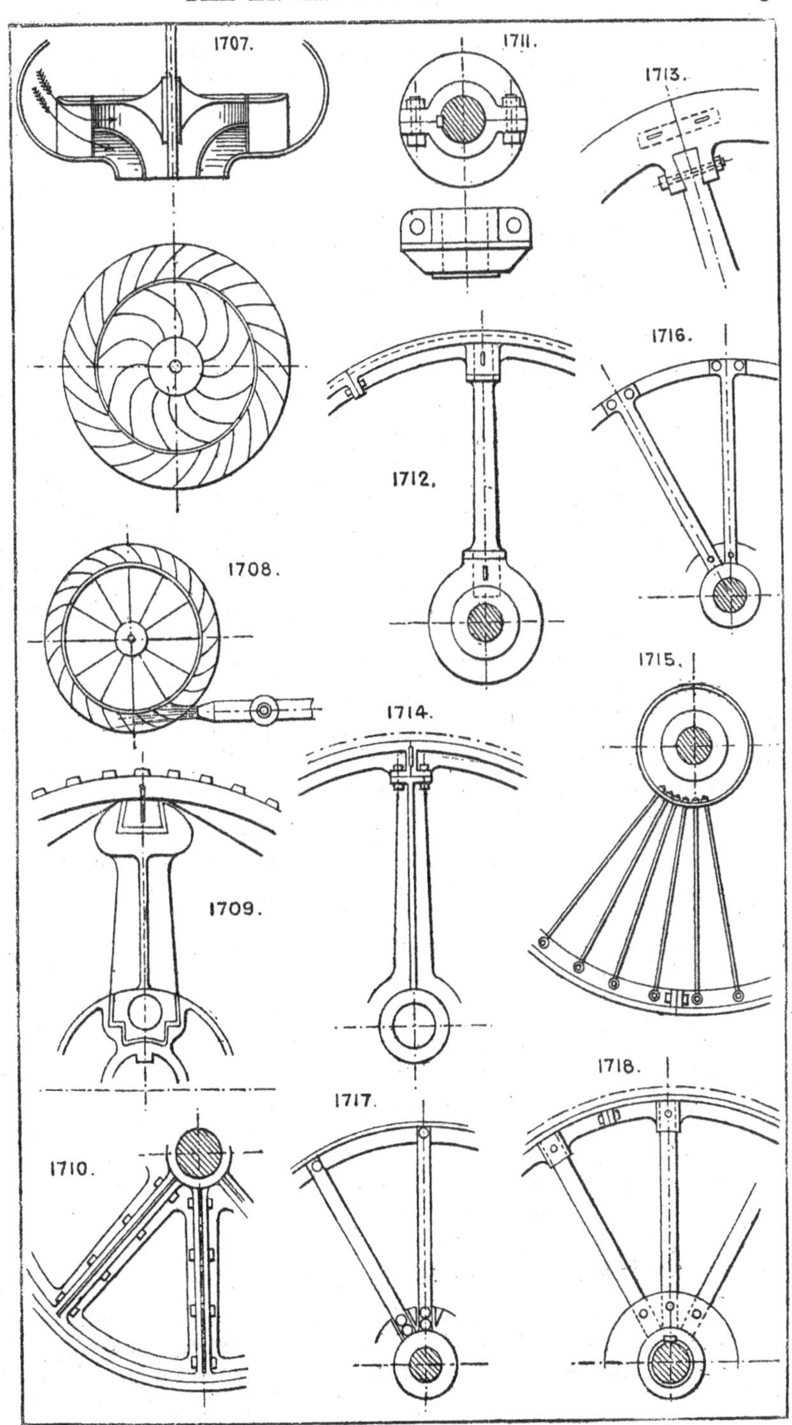
1707.
1711.
1713.
1716.
1712.
1708.
1715.
1714.
1709.
1718.
1717.
1710.

1719. **Railway wheel**: boss of cast iron with wrought-iron arms cast in. The rim is of rolled iron or steel riveted on. There are numerous methods of fastening the tyres to arms, detailed in the *Engineer*, July 23rd, 1880.

1720. **Spring rim split pulley.**

1721. **Segment fly-wheel,** with long radial bolts to secure the rim, arms, and boss together.

1722. **Large centre boss,** with rim segments bolted to it.

1723. **Wheel in halves**; the boss is held together by two bolts acting as cotters.

Section 92.—WEIGHING, MEASURING, INDICATING PRESSURES, &c.

1724. **Weighing by a beam with equal arms.** Weights A = package B.

1725. **Weighing by a beam with unequal arms.** Weight A constant; leverage of ditto variable by shifting it along the graduated arm of lever.

1726. **Graduated measuring vessel.**

1727. **Similar principle applied by compound levers with unequal arms.** The table is supported on four points on the arms of levers loosely jointed together in the centre; one lever is extended and coupled by a rod to a graduated lever with sliding weight. Knife edges are used for bearings for all weighing machines by leverage. See No. 958. This construction is the basis of most of the ordinary weighing machines in use.

1728. **Duckham's patent hydraulic weighing machine.** The article to be weighed is suspended from the hook, and exerts a pressure on the ram. The corresponding pressure on the liquid (usually oil or glycerine) is indicated on the pressure gauge, which is graduated to show the weight.

1729. **Spring balance.**

1730. **Appliance for indicating depth of water** in a cistern by an air bell and pipe connected to a U water gauge. The pressure on the air in the bell varies with the depth or head of water above it, and is indicated on the gauge. A modification of this is employed for sounding at sea.

Weights of substances may be ascertained by their displacement in water or mercury, or by supporting the weighing scale on a free piston resting on an ascertained area of water or mercury, the pressure produced being indicated by a gauge.

1731. **Micrometer gauge.**

1732. **Radial arm weighing machine.**

1733. **Small weighing device,** depending upon the angle the card assumes in respect of the vertical pointer, which is on a free pivot.

1734. **Automatic measuring or weighing device.** The material fills one compartment until it overbalances, when it falls and empties itself; the material then fills the other compartment, and so on.

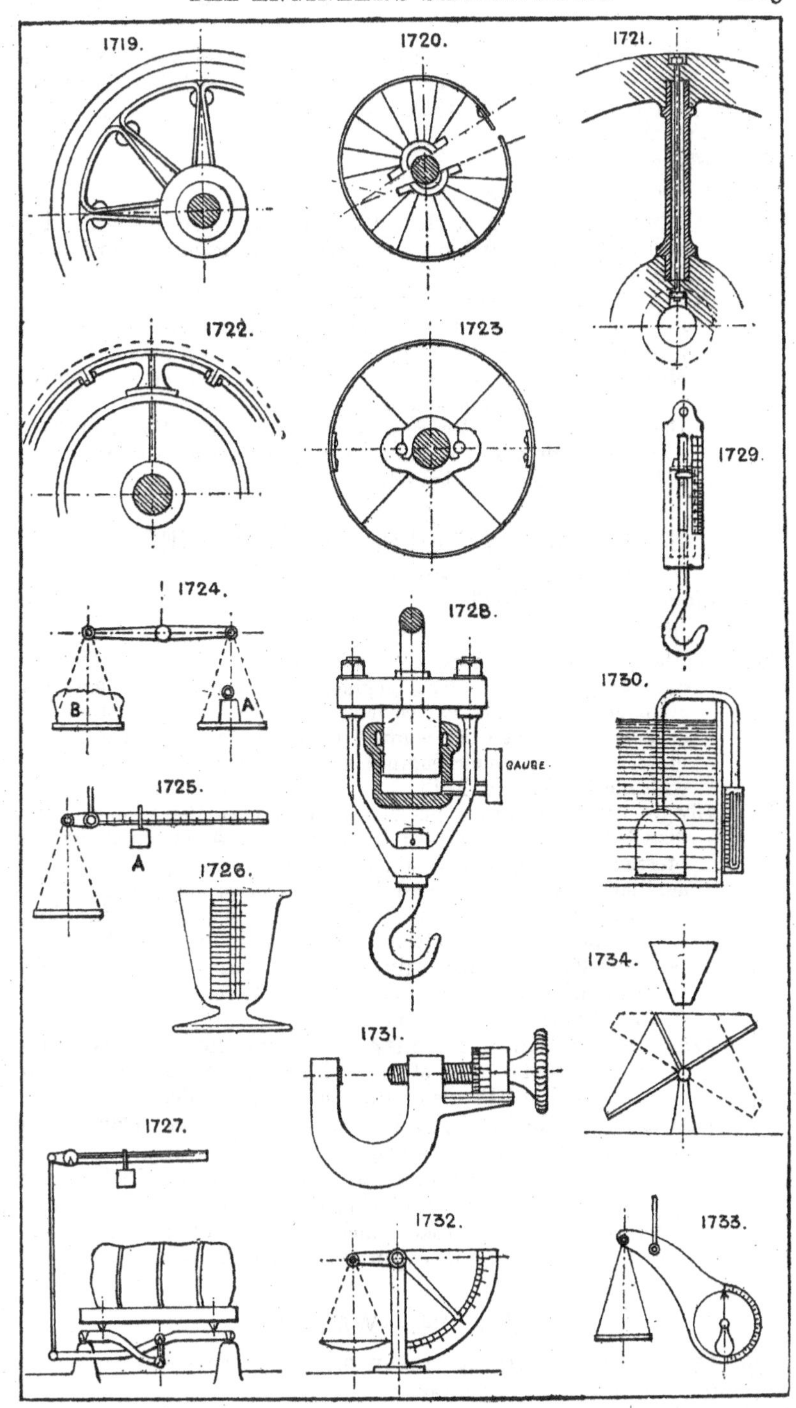
1719.
1720.
1721.
1722.
1723
1729.
1724.
B
A
1728.
1730.
GAUGE
1725.
A
1726.
1734.
1731.
1727.
1732.
1733.

1735. **Wet gas meter.** The gas enters at the centre, and as the compartments fill they rise out of the water, the gas being discharged at the outer ports into the casing.

1736. **Measuring wheel.**

1737. **Measuring wheel.**

1738. **Double slide measurer.**

1739. **Automatic tipping scale.** When full, to equal the weight, it falls and tips by striking a fixed stop; the scale then turns over and returns to its position, and is refilled.

1740. **An ordinary piston and cylinder** are often employed to measure liquids, and fitted with a reversing valve on the same principle as Nos. 1026, 1027, and 1741. See also note to Sec. 93.

Most of the rotary devices (see Section 75) have been employed as meters for liquids and gases. See No. 1692.

Dry gas meters usually employ an expanding bellows, or light piston, with a self-reversing valve, similar to Nos. 1299 and 1026. See also Section 44.

Section 93.—WATER-PRESSURE ENGINES.

See also Section 56.

Hydraulic ram. See No. 1025.

Robinet. See No. 1026.

Two- or three-cylinder engines, with slide valves operated either by eccentrics as in the steam engine, or by the oscillation of the cylinder. The slide valves have no lap or lead; there is no cushioning except what is given by an air vessel on the supply pipe. Three-cylinder engines are usually made single acting with rams. See No. 1743.

1741. **Single cylinder engines.** These must have the slide or other distributing valve reversed by pistons A, A, operated by the pressure of the supply water. This is usually done by an auxiliary valve B, reversed by the main piston rod C. This valve admits the pressure water to the pistons A, A, which reverse the main slide valve. See No. 1506. See note below.

1742. **Mode of working an underground pumping engine** by water cylinders above ground, connected to those below by pipes A, A. B is the suction, C the delivery.

1743. **One, two, or three cylinder water engine.** The ports are in the segmental base of the cylinder, have no lap, and are opened and closed by the oscillation of the cylinder.

1744. **Circular oscillating cylinder,** in a case which opens and closes its ports by its own oscillation.

In lieu of the weighted lever valve gear for single cylinder water pressure engines, the engine may be arranged to compress a spring during the stroke, which at the end of the stroke shall be released, and by its expansion reverse the valve.

Section 94.—WASHING.

1745. **Cylindrical revolving screen washer,** for roots, &c.

1746. **Tub and paddle washer.**

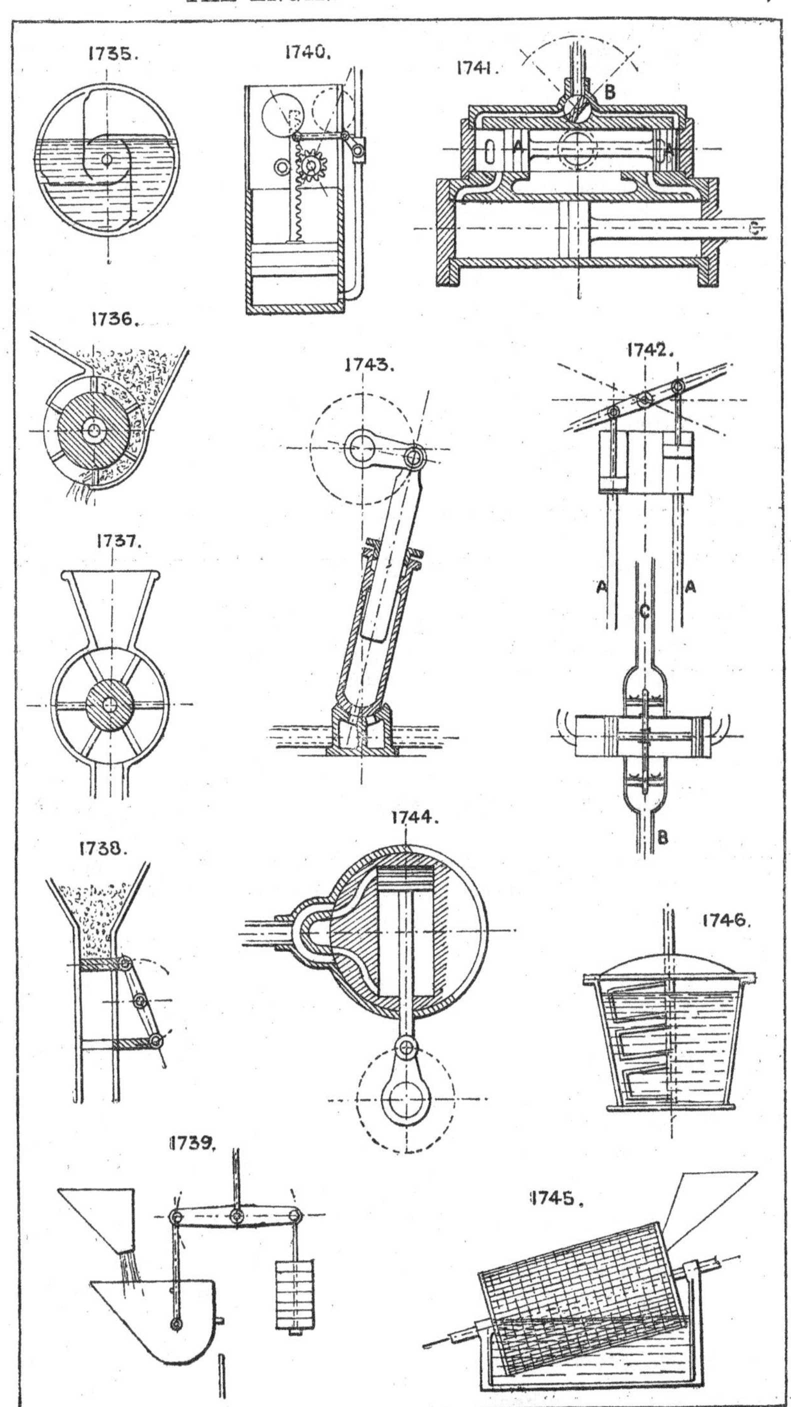
1735.
1740.
1741.
B
A
A
C
1736.
1743.
1742.
1737.
A
A
C
B
1738.
1744.
1746.
1739.
1745.

1747. **Coal washer.** The water is kept in motion up and down through the screen A by a cylinder and piston B; the mud sinks to C and the washed coal passes over to D. Both are removed continuously by elevators or worms. See Section 57.

For washing ores sloping screens either plain or perforated are often used, a stream of water being kept flowing over the ore, which is kept in motion. See Nos. 1266 and 477, also Cylindrical Revolving Screens, as No. 1262.

1748. **Cylindrical perforated drum,** with internal fixed spiral flange which causes the material to travel at a fixed rate of motion. The cylinder may be revolved in a water trough as No. 1745, or water may be fed in with the material and the casing be unperforated.

1749. **A contrivance to keep a continuous circulation** in a boiling tub or copper in which clothes, &c., are washed. The hot water from the bottom rises up the tin tube, and is discharged on the surface.

1750. **Corrugated rollers washing device, for fabrics.**

1751. **Water trough and dipping band,** for washing cloths, wool, paper, &c.

Domestic washers comprise, besides the ordinary tub and dolly, washing boards, having corrugated surfaces; rocking and revolving boxes, having a churn-like motion. Brushes also are sometimes used.

Section 95.—WINDMILLS AND FEATHERING WHEELS.

1752. **Feathering paddle wheel.** Each float has a bracket and pin at back, with connecting rod to a common eccentric (fixed), through which the shaft revolves.

1753. **Spiral vane or cowl,** for chimney top. Used to drive a vertical worm inside the chimney cap to maintain an upward draught, by employing the wind as a motive power.

1754. **Windmill sails,** with angular adjustment by a sliding device on the shaft.

1755. **Feathering horizontal windmill.** Each float is hinged a little out of centre to the arms, so that the pressure of wind (see arrow) turns the floats to the positions in the sketch as they revolve.

1756. **Hollow semi-spherical cup windmill** or motor.

1757. **Wind motor,** with curved vanes. These last two revolve in the direction of the arrows, because the wind has more hold upon the hollow sides of the cups and vanes than on the convex side.

1758. **Self-feathering wind wheel.**

1759. **Spiral wind wheel.**

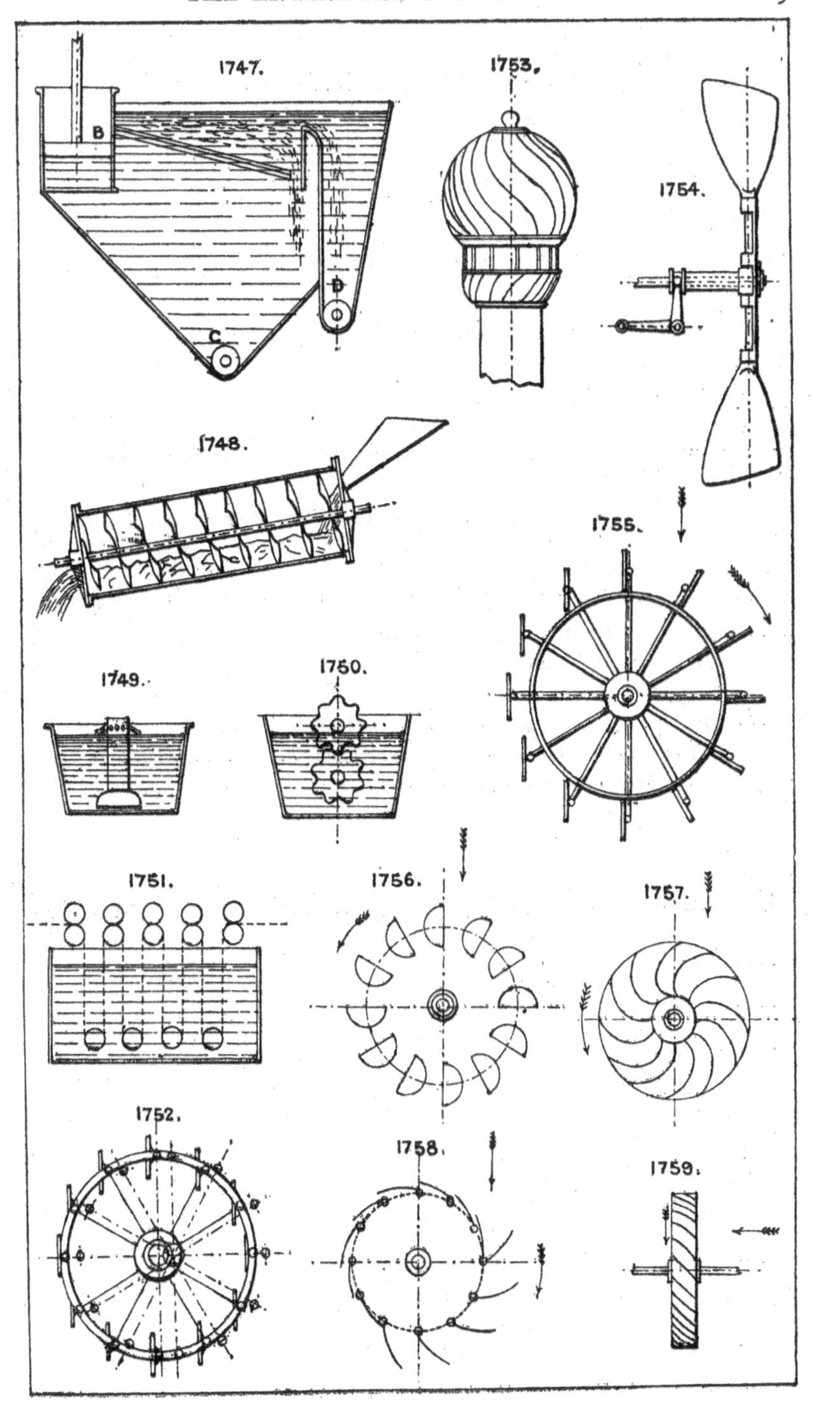
1747.
B
C
D
1748.
1749.
1750.
1751.
1752.
1753.
1754.
1755.
1756.
1757.
1758.
1759.

Section 96.—WINDING APPARATUS.

1760. **Barrel or drum**, for wire, &c.

1761. **Winding barrel**, for cranes, winches, &c.

1762. **Fusee barrel.** See No. 1602.

1763. **Grooved barrel**, for chain. Prevents the chain riding as it coils.

1764. **Hexagon frame winder**, for yarn, &c.

1765. **Spool.**

1766. **Card winder.**

1767 & 1768. **Bobbins.** There are a great many patterns in use for various trades.

1769. **Appliance for winding bobbins** of cotton, and other machinery. The bobbin is stationary and the flyer revolves, the thread passing up its centre and down one arm through the eye, which has an up and down feed motion to wind the thread on evenly.

1770. **Mode of feeding the thread** on to the spindles to cause it to coil evenly by an oscillating arm and pin over which the thread passes.

1771. **Drum**, for flat rope or chain wound upon itself.

For winding engines and winches, see Nos. 1222 to 1226.

See also Rope Gear, Section 66.

Section 97.—HANDLES, &c., FOR VARIOUS PURPOSES.

1772. **Knob handle.**

1773. **Loop handle**, hinged.

1774. **Loop handle**, fixed.

1775. **T handle.**

1776. **Plain handle.**

1777. **Sash lift** or drawer handle.

1778. **Hand bar.**

1779. **Swing door handle.**

1780. **Sunk or flush loop handle.**

1781. **Hinged lifting levers.**

1782. **Bent handle**, for radial motion.

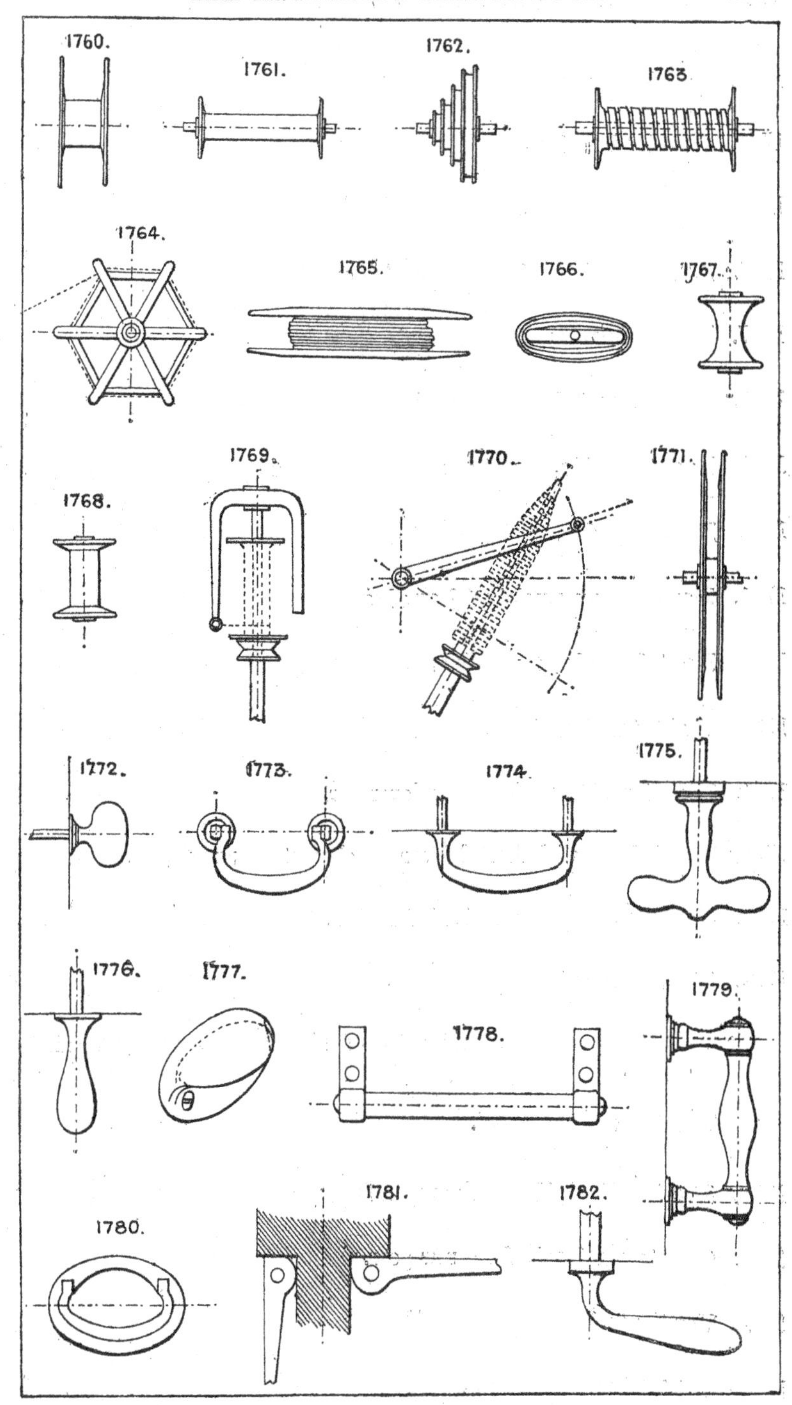
1760.
1761.
1762.
1763
1764.
1765.
1766.
1767.
1768.
1769.
1770.
1771.
1772.
1773.
1774.
1775.
1776.
1777.
1778.
1779.
1780.
1781.
1782.

1783. **Hand wheel.**

1784. **Cranked T handle.**

1785. **Capstan wheel.**

1786. **Bow or lifting handle,** for ladles, buckets, &c.

1787. **T bar handle,** for two hands.

1788. **Cross hand lever,** four, six, or eight arms.

1789. **Loop handle,** sometimes cast into a casting.

1790. **Ring handle.**

1791. **Double bar pushing handles.**

1792. **Bent handle,** for radial motion.

1793. **Weighted handle.**

1794. **Vice handle,** with sliding lever bar.

1795. **Hand bar,** with forked lever attachment for pumps, &c.

1796. **S lever double cranked handle.**

1797. **Stirrup handle.**

1798. **T lifting handle or key,** for opening flush doors or manhole covers.

1799. **Thumb screw head.**

1800. **Straight handle,** with suspending eye.

1801. **Capstan wheel,** for screw gear.

1802. **Ventilated twisted handle.**

1803 & 1804. **Loop handles.**

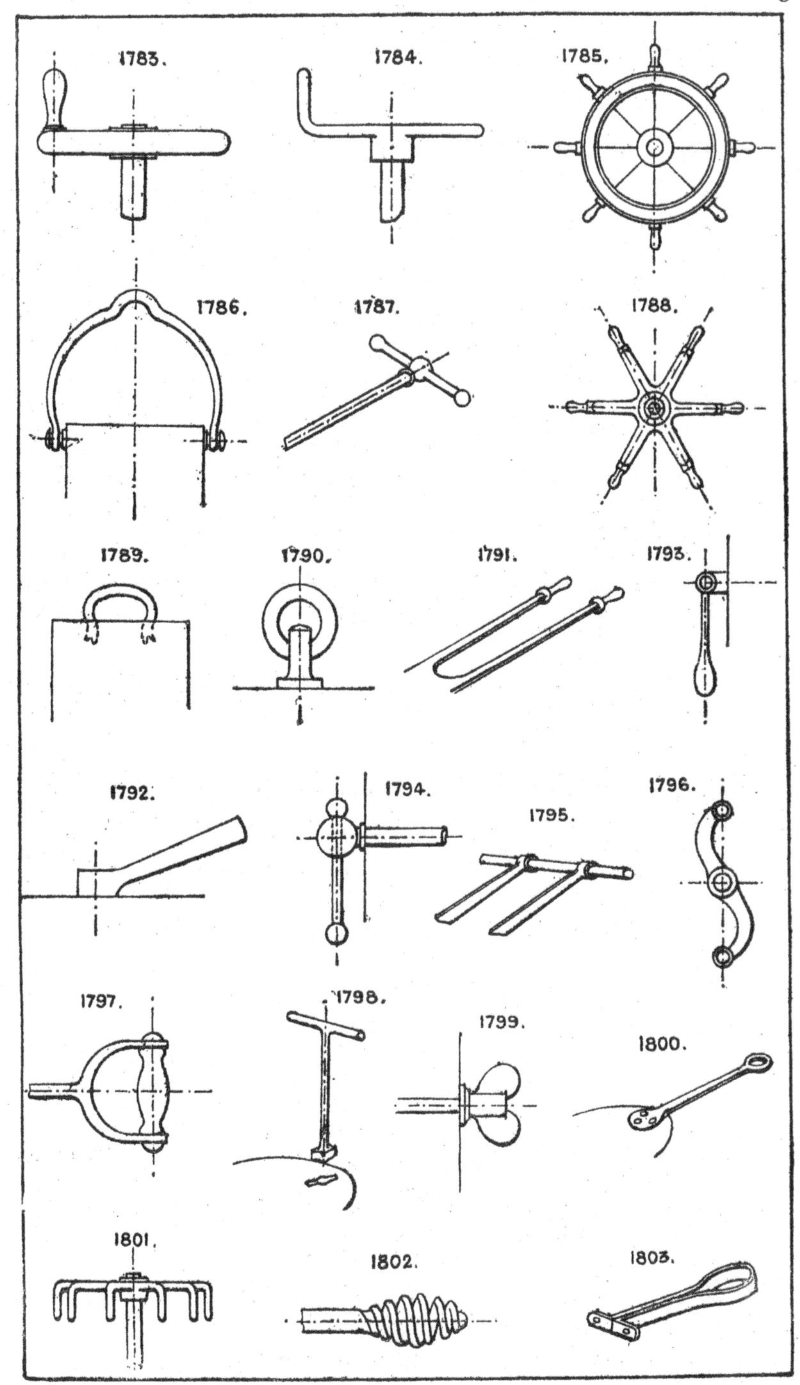
1783.
1784.
1785.
1786.
1787.
1788.
1789.
1790.
1791.
1793.
1792.
1794.
1795.
1796.
1797.
1798.
1799.
1800.
1801.
1802.
1803.

1805. **Spring lock lever handle.**
1806. **"Coffee pot" handle.**
See also Section 48.

Section 98.—APPARATUS FOR DRAWING CURVES.

See Gearing, Section 40; Ellipsograph, Section 34.

1807. **Cyclograph,** for describing arcs, the chord and versed sine being given. Fix pins at A and B, fasten together at C two slips of wood, hold the pencil at D, and move the slips round, keeping them against the pins. See also No. 11.

1808. **Hyperbolagraph.** Height and focus are given as for 1809, and string fixed to B, and focus A; the arm is pivoted at C, and the pencil used as described for No. 1809.

1809. **Parabolagraph.** The height of parabolic curve H and focus A are given. A string is fastened to end of set square at B, reaching to C, and the other end fixed to a pin in the focus A. A pencil held in the loop and kept against edge of set square as it is moved to left or right will describe the parabolic curve.

1810. **Cycloidograph** describes the hypo- or epi-cycloid. A modification of this is used to draw the curves of the teeth of wheels.

Pentagraph, for reducing or enlarging outline drawings, No. 1924.

Helicograph, to describe a regular helix by a central fixed bevil wheel which drives the radial bevil wheel screw and scribing pencil, No. 1925.

A simple helicograph, with radial screw and roller nut, which travels along the screw as the apparatus is revolved on its centre pin, No. 1926.

Section 99.—MATERIALS EMPLOYED IN CONSTRUCTION.

The following memoranda relate only to such materials as are required in connection with machinery or mechanical constructions, and are intended to supply particulars of the dimensions of the manufactured or raw material, giving the sections manufactured and the limits as to size available for incorporation in any design under consideration.

Rolled iron and steel bars are manufactured as below:—

1811. **Rounds,** from $\frac{3}{16}''$ to $7\frac{3}{4}''$ diameter, and up to 18′ long.

1812. **Squares,** from $\frac{3}{16}''$ to 6″ square, and up to 18′ long.

1813. **Flats,** from $\frac{1}{2}''$ to 14″ wide, and up to 18′ long.

1814, 1815, 1816, 1817, 1818, & 1819. **L iron sections** are made from $\frac{3}{4}'' \times \frac{3}{4}''$ up to $14'' \times 3\frac{3}{4}''$, or to $12\frac{1}{2}''$ united inches, with equal or unequal flanges, and up to 30′ long; but the acute, obtuse, and round angled sections are not usually stocked.

1820 & 1821. **T irons,** from $1'' \times 1''$ up to 12 united inches, or to $9'' \times 4''$, and up to 30′ long.

1822. **Rolled girder iron,** from 3″ deep to 20″ deep × 10″ flanges, and to 36′ long, in hundreds of sections.

1823. **Zore girders,** from 3″ to 8″ deep, and to 24′ long.

1824. **Channel iron,** from $\frac{3}{4}''$ to 12″ wide, and to 25′ long.

1825. **Convex iron,** from 1″ to 6″ wide, and up to 20′ long.

1826. **Cope iron,** from 1″ to 4″ wide, and to 20′ long.

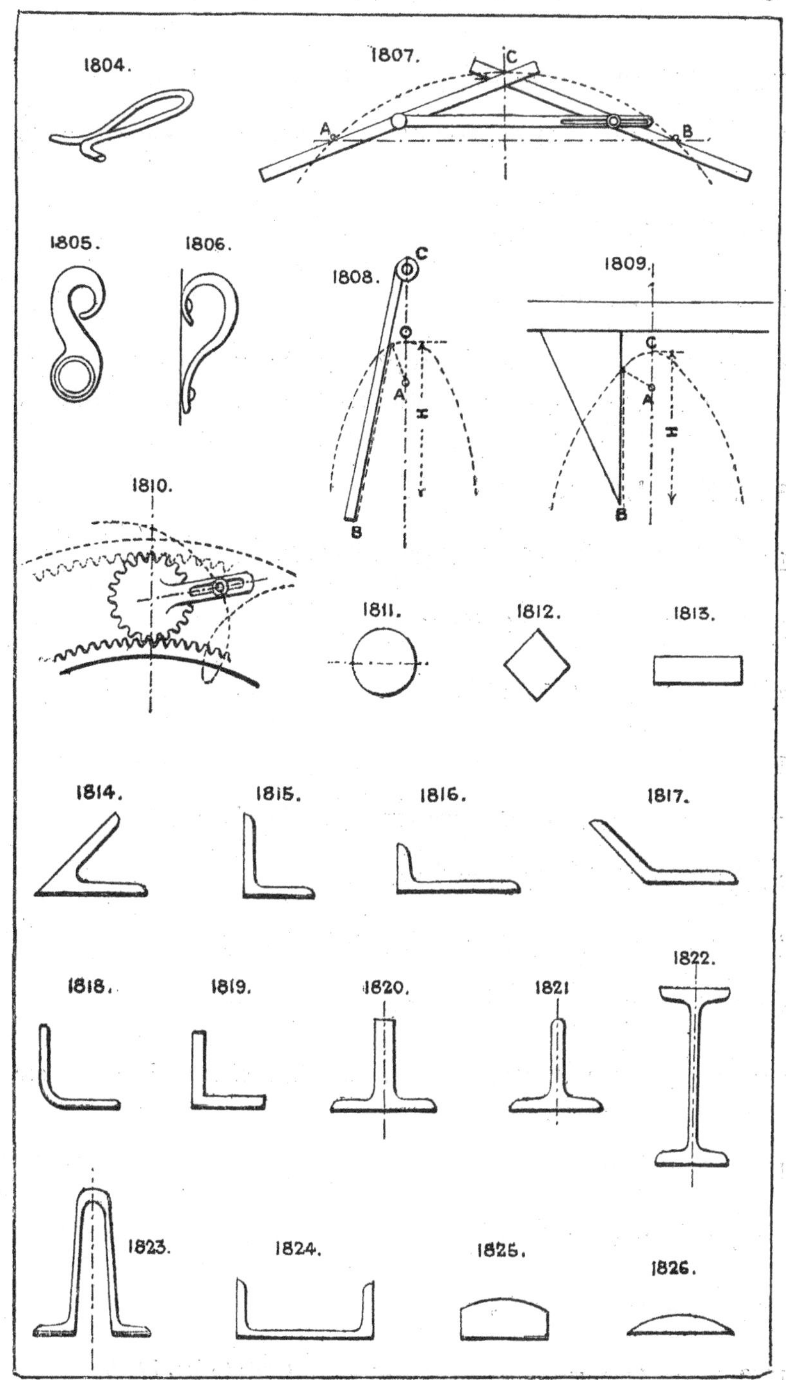
1804.
1807.
C
A
B
1805.
1806.
1808.
C
A
H
B
1809.
C
A
H
B
1810.
1811.
1812.
1813.
1814.
1815.
1816.
1817.
1818.
1819.
1820.
1821
1822.
1823.
1824.
1825.
1826.

1827. **Half-round iron,** from $\frac{1}{2}''$ to $4''$ wide, and to $20'$ long.

1828. **Funnel ring iron,** from $3\frac{1}{2}'' \times \frac{3}{16}''$ to $8'' \times \frac{9}{16}''$ wide, and up to $18'$ long.

1829. **Jackstay iron.**

1830. **Hollow cope iron.**

1831, 1832, & 1838. **Rail sections** (see Section 73), usually made in $18'$ to $30'$ lengths, and numerous sections of from 22 lbs. to 84 lbs. per yard.

1833. **Bulb L iron.**

1834. **Deck beam or bulb T iron,** up to $16'' \times 6''$.

1835. **Bulb L iron,** up to $10'' \times 4''$.

1836. **Bulb iron,** to $13''$ wide.

1837. **Pile iron.**

1839, 1840, & 1841. **Flush tram rails,** $18'$ to $30'$ long.

1842, 1843, & 1849. **Fire bar iron.**

1844. **Double L iron,** $\frac{1}{2}'' \times 1'' \times \frac{1}{2}''$ to $5'' \times 5'' \times \frac{1}{2}''$.

1845. **Cross iron.**

1846, 1847, & 1853. **Sash bar iron.** Hundreds of special sections are manufactured.

1848. **Bevil edge iron.**

1850. **Octagon bar iron.**

1851. **Hexagon bar iron.**

1852. **Tyre iron,** made in many sections. See note to No. 1719.

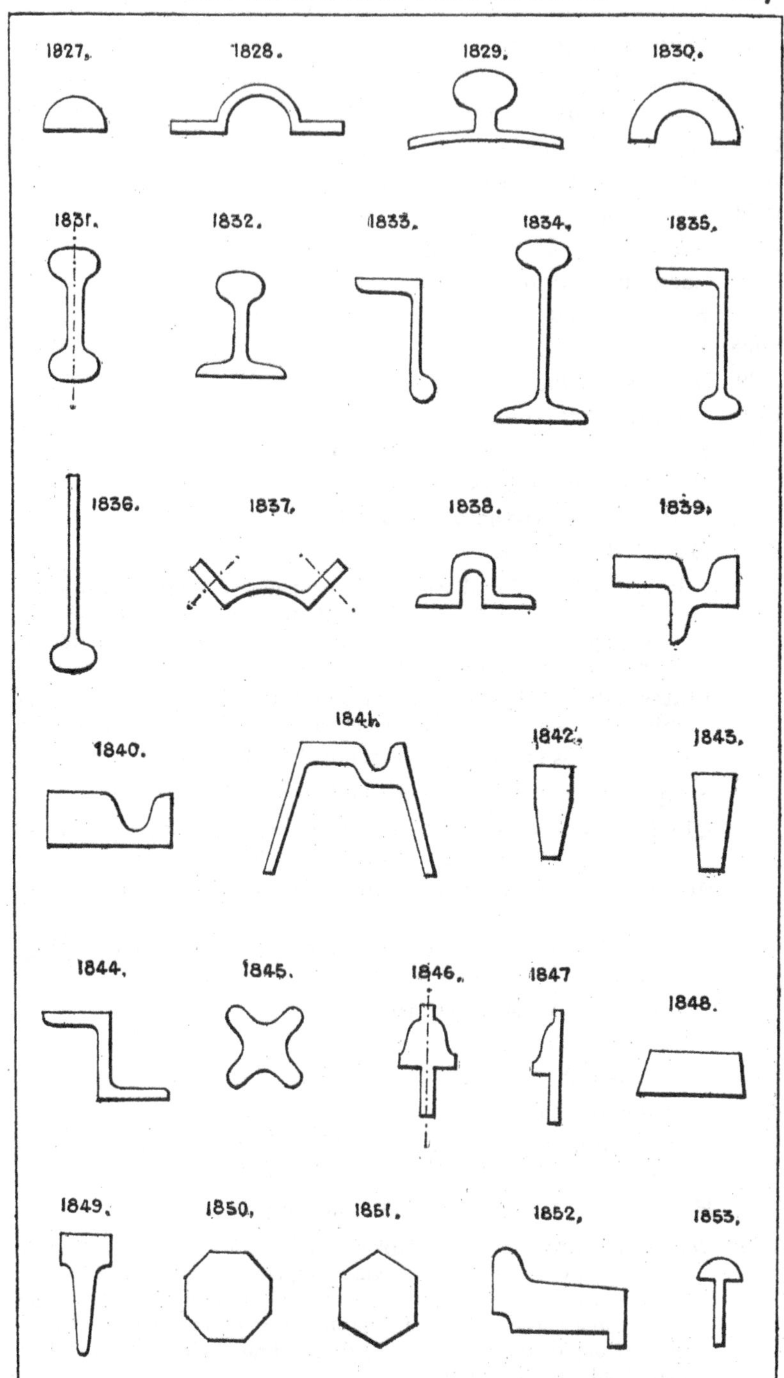
1827.
1828.
1829.
1830.
1831.
1832.
1833.
1834.
1835.
1836.
1837.
1838.
1839.
1840.
1841.
1842.
1843.
1844.
1845.
1846.
1847
1848.
1849.
1850.
1851.
1852.
1853.

1855. **Bevilled flat iron.**

1856. **Trough iron.** Used for bridge flooring, fire-proof floors, &c.

1857. **Double convex iron.**

1858 & 1859. **Tramplate iron.**

1860 & 1861. **Chair or sleeper iron.**

1862. **Oval iron.**

1863, 1864, & 1865. **Round edged flats.**

1866. **Segment round iron.**

1867. **Round edged convex iron.**

1868. **Bevilled flat iron.**

1869. **Bevil edge flat iron.**

1870. **Bevilled flat iron.**

1871. **Round edged hollow convex iron.**

1872. **Taper edged hollow convex iron.**

1873. **Boiler tube expansion ring iron.**

1874. **Moulded flat bar.**

In addition to the above, iron ornamental mouldings are rolled with moulded and relief ornaments in bars, from $\frac{5}{8}''$ to $2\frac{3}{4}''$ wide, and up to 16′ or 18′ long. Also plain mouldings similar in sections to those used in joinery.

Plates (iron and steel) are manufactured from $\frac{1}{8}''$ to $\frac{3}{4}''$ thick ordinary. Thicker plates are rolled to order up to 20″ thick.

Stocked sizes of ordinary plates are 4′ × 2′ up to 14′ × 4′ 6″.

Strips from 7″ to 22″ wide, and up to 30′ long.

Chequered plates, with diamond, oval or square recessed patterns, are made 6′ × 2′ up to 8′ × 3′ 6″.

Sheets, plain, in thicknesses from No. 10 w.g. to No. 36 w.g., and from 6′ × 2′ to 10′ × 4′.

Corrugated sheets, plain or galvanised, from No. 16 to No. 26 w.g., and from 6′ × 2′ to 9′ × 2′.

Tinned sheets, same as above.

Cold rolled sheets ,, ,,

Planished sheets ,, ,,

Lead-coated sheets ,, ,,

Tin plates, terne plates, 14″ × 20″, 17″ × $12\frac{1}{2}''$, 15″ × 11″, 14″ × 10″, 24″ × 20″, 28″ × 10″, 28″ × 20″.

Hoops, from $\frac{5}{8}''$ to 7″ wide, and from No. 8 to No. 24 w.g.

1875. **Wire; sections manufactured** in hard iron, soft iron, soft steel, hard steel, tempered steel, piano wire, covered wire (wound with either cotton, silk, guttapercha, flax, &c.), or copper wire. Also brass, copper, lead, zinc, and other metal wire, hard or soft; tinned iron wire, galvanised iron wire, tinned brass wire, coppered iron wire, lead-coated iron wire.

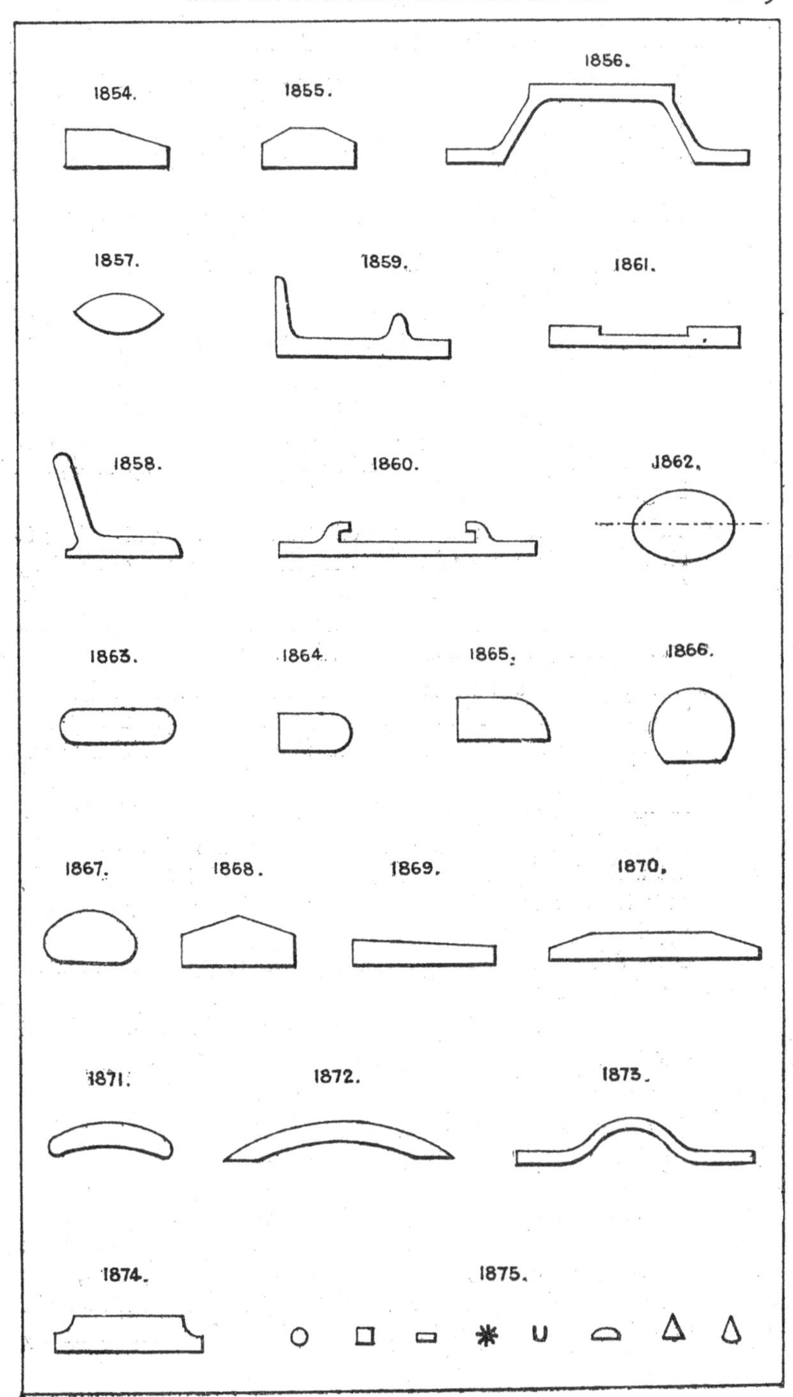
1854.
1855.
1856.
1857.
1859.
1861.
1858.
1860.
1862.
1863.
1864.
1865.
1866.
1867.
1868.
1869.
1870.
1871.
1872.
1873.
1874.
1875.

Pipes (see Section 57) **and tubes** of wrought iron, either butt or lap welded, or solid drawn, are made in four qualities or strengths:—1. Gas tube; 2. Steam or water tube; 3. Boiler flue tube; 4. Hydraulic tube. These are manufactured from ¼″ to 3″ internal diameters; boiler flue tubes to 9″ diameter, but much larger sizes can be made to order.

Solid drawn steel tubes are made up to 10″ diameter; larger sizes are made to order.

Special steel or wrought iron pipes, flanged with L iron, are made up to 4′ diameter with welded joints, and welded steel or wrought iron socket and spigot pipes up to 24″ diameter.

Cast iron pipes are made in the following strengths:—Rain-water pipes, hot-water pipes, gas mains, water mains, hydraulic mains for high pressure, and the thicknesses of metal vary according to the pressures. Diameters from 1½″ up to 4′, and lengths usually 6′ and 9′. See Section 57.

Castings are made in cast iron of various mixtures, according to strength, toughness, or hardness required, and of any weight up to 20 tons. Chilled iron castings are made for hard wear, as in crusher rolls, &c., but cannot be machined; they are usually ground smooth by a grindstone or emery wheel.

Steel castings are made in either Bessemer, Siemens-Martin, Thomas-Gilchrist, or in crucible steel, the latter being most relied upon. They require annealing to soften them sufficiently for machining, are almost invariably "blown" or honeycombed, and rarely homogeneous, or twice alike from the same pattern or cast.

Wrought-iron castings, Mitis metal, &c., are also obtainable, but malleable cast iron castings are most relied upon for toughness, the process having now attained great perfection, but is not applicable to very thick castings.

Pressed iron on steel forgings of simple forms are now obtainable at low prices.

Forgings in wrought iron and steel can now be made to almost any size, shape, and weight, and are replacing many structures formerly made of cast iron or built up.

Other metals employed are copper, brass, tin, zinc, phosphor-bronze, lead, antimony, bismuth, pewter, Muntz metal, aluminium, sodium, potassium, platinum, gold, silver, nickel, and a great variety of the bronzes, which are valuable compounds varying in tenacity and hardness from the hardest steel to that of soft copper. Most of the above are manufactured into wire, sheets, tubes, rods, &c., and can in addition be cast into any form from a crucible. Copper can be forged but not welded; joints in it are generally brazed or soldered.

Other materials employed comprise—

Timber. Yellow, white, and red pine in logs, deals, and battens; logs, up to about 3′ diameter by 35′ to 40′ long; deals, 9″, 10″, and 11″ wide, and from $1\frac{1}{2}$″ to 4″ thick—a few wide deals are imported up to 22″ wide—spruce and fir, sycamore, pear tree, willow, poplar, &c. The following table gives a list of woods and their applications:—

TABULAR STATEMENT OF THE WOODS COMMONLY IN USE IN GREAT BRITAIN.

For Building.

Ship-building.—Cedars, deals, elms, firs, larches, locust, oaks, &c., &c.

Wet works, as piles, foundations, &c.—Alder, beech, elm, oak, plane-tree, white cedar.

House carpentry.—Deals, oaks, pines, sweet chestnut.

For Machinery and Mill-work.

Frames, &c.—Ash, beech, birch, deals, elm, mahogany, oak, pines.

Rollers, &c.—Box, lignum vitæ, mahogany.

Teeth of wheels, &c.—Crab-tree, hornbeam, locust.

Foundry patterns.—Alder, deal, mahogany, pine.

For Turnery.

Common wood for toys (softest).—Alder, beech (small), birch (small), sallow, willow.

Best woods for Tunbridge ware.—Holly, horse chestnut, sycamore (white woods); apple-tree, pear-tree, plum-tree (brown woods).

Hardest English woods.—Beech (large), box, elm, oak, walnut.

For Furniture.

Common furniture and inside works.—Beech, birch, cedars, cherry-tree, deal, pines.

Best furniture.—Amboyna, black ebony, cherry-tree, Coromandel, mahogany, maple, oak (various kinds), rose-wood, satin-wood, sandal-wood, sweet chestnut, sweet cedar, tulip-wood, walnut, zebra-wood.

Foreign hard woods, several of which are only used for ornamental turnery.—

1. Amboyna.	13. Greenheart.	25. Peruvian.
2. Beef-wood.	14. Grenadillo.	26. Princes-wood.
3. Black Bot. B. wood.	15. Iron-wood.	27. Purple-wood.
4. Black ebony.	16. King-wood.	28. Red sanders.
5. Box-wood.	17. Lignum vitæ.	29. Rosetta.
6. Brazil-wood.	18. Locust.	30. Rose-wood.
7. Braziletto.	19. Mahogany.	31. Sandal-wood.
8. Bullet-wood.	20. Maple.	32. Satin-wood.
9. Cam-wood.	21. Mustaiba.	33. Snake-wood.
10. Cocoa-wood.	22. Olive-tree & root.	34. Tulip-wood.
11. Coromandel.	23. Palmyra.	35. Yacca-wood.
12. Green ebony.	24. Partridge-wood.	36. Zebra-wood.

Nos. 3, 8, 16, 33, and 34 are frequently scarce.

Nos. 3, 5, 8, 9, 10 are generally close, hard, even tinted, and the more proper for eccentric turning, but others may also be employed.

Nos. 4, 5, 10, 12, 14, 17, 18, 19, 30, 32 are generally abundant and extensively used. All the woods may be used for plain turning.

Miscellaneous Properties.

Elasticity.—Ash, hazel, hickory, lance-wood, sweet chestnut (small), snake-wood, yew.

Inelasticity and toughness.—Beech, elm, lignum vitæ, oak, walnut.

Even grain, proper for carving.—Lime-tree, pear-tree, pine.

Durability in dry works.—Cedar, oak, poplar, sweet chestnut, yellow deal.

Colouring matter (red dyes).—Brazil, braziletto, cam-wood, log-wood Nicaragua, red sanders, sapan-wood.

Colouring matter (green dye).—Green ebony.

Colouring matter (yellow dyes).—Fustic, zantes.

Scent.—Camphor wood, cedar, rose-wood, sandal-wood, satin-wood, sassafras.

Indiarubber, manufactured into sheets, with or without canvas insertion of single, double, or treble thickness, up to 36″ wide and to ½″ thick; cord ⅛″ to 1″ diameter; tubes, plain, or with canvas insertion or wire coiled inside or outside, from ¼″ to 4″ bore, usually in 30′ and 60′ lengths. Washers, rings, rollers, strips, belts, and moulded articles of every form.

Guttapercha is manufactured into similar articles.

Leather. Most of the varieties are manufactured from the skins of oxen, sheep, goats, deer, horses, dogs, hogs, and seals, and the larger skins are divided into butts, shoulders, cheeks, and bellies, the dimensions depending of course upon the size of the animals. Ox hides are the largest and kid skins the smallest in general use.

For mechanical purposes ox hide, raw or tanned, is chiefly used, as for valves, seatings, belts, piston leathers, &c. Sheep skins can be obtained either strained, half-strained, or unstrained; the first are hard and comparatively stiff, the last-named soft and pliable as cloth. Other soft varieties are goats' skins and chamois leather. There are many imitations of leather, but they are rarely employed in mechanical constructions.

Vulcanised fibre is often used for similar purposes to leather, as for valves, seatings, joints, &c. It is made in two varieties, medium and hard, and in sheets up to 1″ thick.

Ebonite. A hard, black, horny substance, moulded into any required shape.

Papier mâché. Solid paper, moulded from pulp into any required form.

Asbestos, in sheets, cord, packing of various sections, loose fibre, mill-board, &c.

Ivory, from tusks and teeth.

Bone.

Vegetable ivory; nuts about the size of eggs.

Packings for glands, &c., are made of cotton, hemp, and other fibres, asbestos, indiarubber, &c., in round, square, and other sections.

Section 100.—HEATING APPARATUS.

For general purposes this comprises Furnaces, Stoves, Ranges, Ovens, Boilers (see Section 6), Hot-blast, Steam-heated Vessels, Gas Jets, &c., most of which are tolerably well known and in common use.

For special purposes in connection with machinery various heating devices are required, of which steam and gas are those most universally used. Steam tubes or coils may be carried through any fixed or movable part of a machine. Steam-heated surfaces, such as tables, pans, chambers, &c., steam-jacketed cylinders, and similar contrivances, are much used. Gas jets from perforated tubes, which may be shaped to any required position, are also convenient for dry heat and higher temperatures than can be obtained from steam.

Hot irons are sometimes used, shaped to fit a cavity, but of course require to be replaced periodically.

Hot water in pipes or jackets, and hot air in flues are common appliances for warming and drying; with the former its circulation must be provided for, and with the latter, either a forced draught or an upward inclination given to the flues to maintain circulation.

1876. **Gill pipes** for radiating the heat of steam or hot water.

1877. **Gill stove,** on similar principle, presents an extensive surface in contact with the air for radiation of heat.

Section 101. DRAWING AND ROLLING METALS, &c.

1878. **Rolls for bar iron,** grooved to suit the section required, one-half the groove being usually in each roll, and the size and shape of the grooves are graduated down from that of the square billet to the finished bar.

1879. **Grooved rolls** for producing a tapered bar.

1880. **Rollers for turning up** and welding tubes from a flat strip.

1881. **Bending rollers.**

1882. **Rolls for solid tyres,** without a weld.

1883. **Wire drawing apparatus.**

For grips for drawing wire, &c., see Nos. 505, 518. Laths of various sections are drawn through suitable steel dies by a draw bench; the end of the lath is held by a grip tongs and the lath drawn forcibly through the dies (using a lubricant) and afterwards straightened. Rolling does not answer for this kind of work.

The drawing frame used for cotton and other fibres has two, three, or more pairs of rollers; the lower rollers are grooved longitudinally and the upper ones weighted and covered with leather, the lower ones being geared together to drive at proportionate speeds, so that in passing through, the material is stretched between each pair of rollers, the object being to extend and lay all the fibres parallel.

For drawing lead pipes, see No. 1183. Earthenware pipes are made by a similar process.

Section 102.—STRUTS AND TIES.

1884. **Ordinary solid swelled distance rod** with collars, used for compressive strains.

1885. **Similar strut,** but formed of tube with end collars screwed in.

1886. **Double flat-bar cambered strut,** stiffened by distance pieces and bolts.

1887, 1888, 1889, & 1890. **Sections of varieties of the foregoing.**

1891. **Braced strut;** usually of flat bars on edge, riveted together at the intersections.

1892. **Tubular swelled strut,** of plate iron, used for masts, sheer legs, crane jibs, &c.

1893. **Built up strut,** from segmental bars.

1894. **Trussed strut;** the trussing is 90° apart, but may be at any angle; the central bar of course takes the actual thrust, and the truss rods keep it from bending or buckling. See also Nos. 295 to 300, 320.

Ties, for tensile strain only are usually of round iron, flat or other simple section, tube, or even chain, rope, or wire.

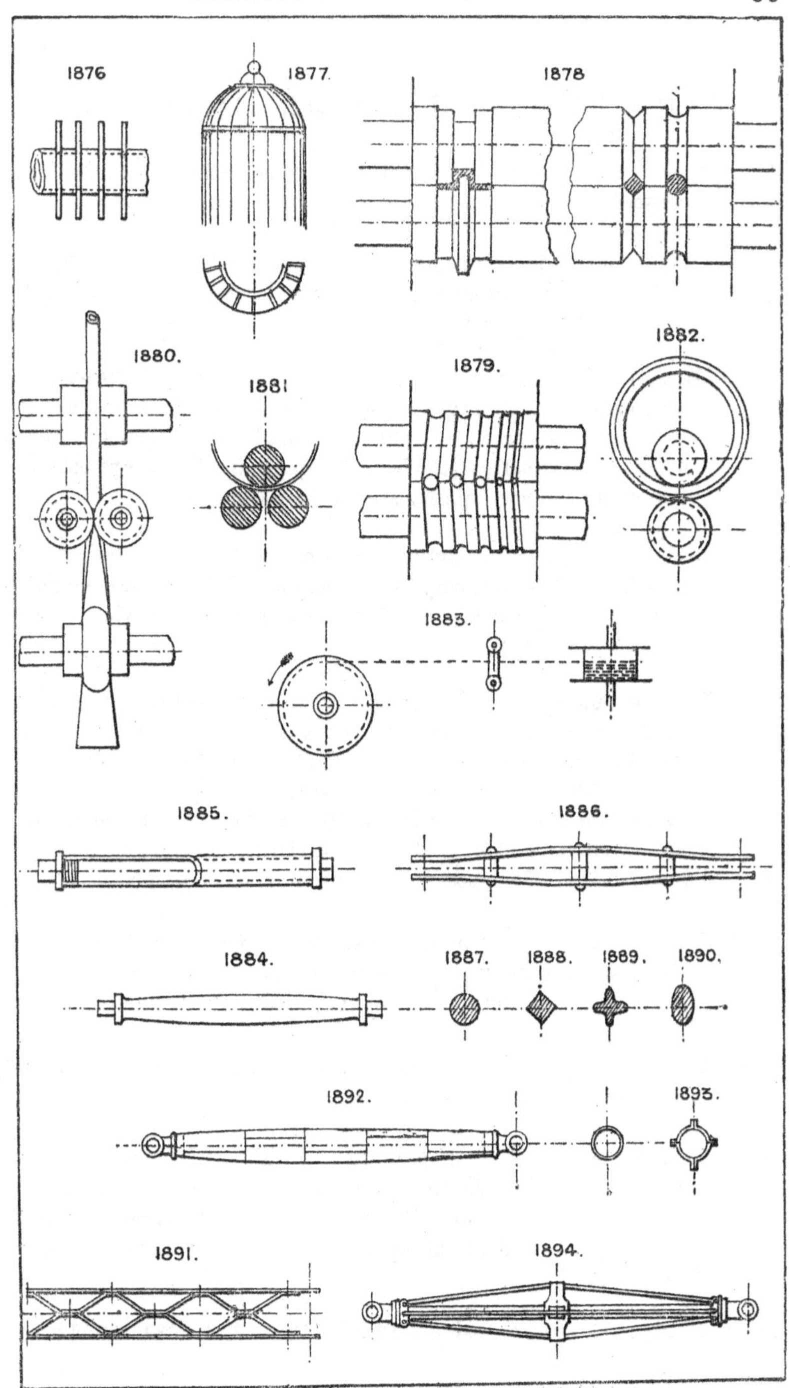
1876
1877.
1878
1880.
1881
1879.
1882.
1883.
1885.
1886.
1884.
1887.
1888.
1889.
1890.
1892.
1893.
1891.
1894.

Section 103.—MARINE ENGINES (TYPES OF).

Many varieties will be found illustrated under Section 32. The following are modern types :—

1895. **Diagonal paddle engines,** for light draught vessels. May of course be either of two or three cylinder type and either high pressure or compound.

1896. **One of the most favourite types of vertical overhead cylinder screw engines,** with half standards and distance rods, one, two or three cylinders, simple or compound. The condenser is usually in the back standards and the pumps behind. Simplicity and accessibility are its chief advantages.

1897. **Stern wheel, side lever engines,** not often required in practice. The ordinary construction of horizontal engines usually accommodates itself for stern wheel driving. See Nos. 575 to 579, &c.

1898. **Double standard vertical overhead cylinder screw engines,** the type commonly adopted for the heavier class of vessels, and frequently made for triple expansion. It is of very rigid construction, but not quite so convenient for accessibility to the working parts as No. 1896. The condenser and pumps are at one side, built into the standard, and the engines are handled from the opposite side or from an elevated platform.

1899. **Overhead cylinder and distance rod type,** the lightest and simplest form in use for small engines. Every part is easily seen and got at, and the top weight is reduced to a minimum.

1900. **Is a variety of No. 1896,** with tandem cylinders and two cranks for triple expansion.

1901. **Is also a variety of No. 1898,** with tandem cylinders for triple expansion. In this plan the intermediate stuffing box is got rid of by using two piston rods to the lower cylinder, coupled to the piston rod of the upper or high-pressure cylinder by a crosshead.

1902. **Compound overhead standard engines,** for twin screws.

1903. **Diagonal twin screw engines.**

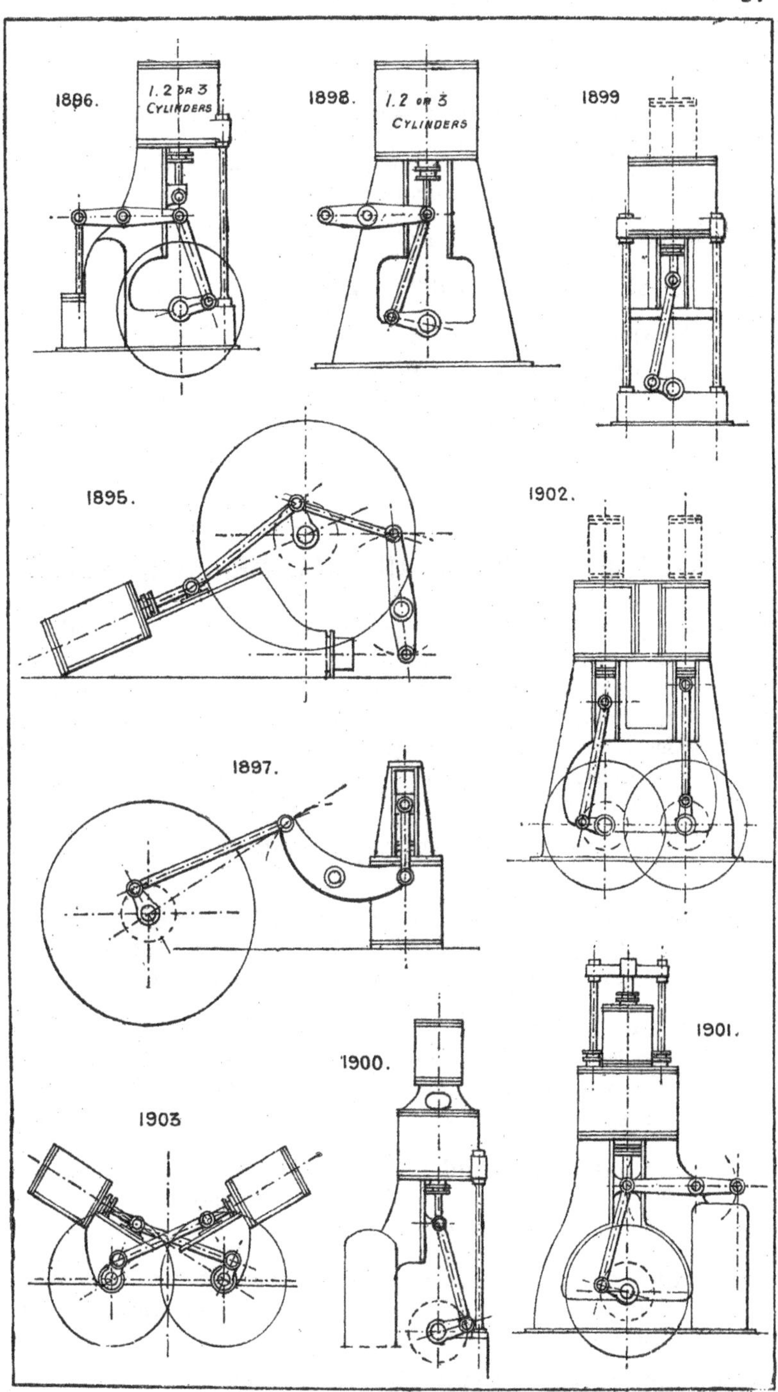
1896.
1. 2 or 3 CYLINDERS
1898.
1. 2 or 3 CYLINDERS
1899
1895.
1902.
1897.
1900.
1901.
1903

1904. **Horizontal twin screw engines.**

1905. **Plan of cylinders** as usually employed for No. 1902.

1906. **Oscillating paddle engines,** sometimes made with cylinders at 90° apart and a single crank, as No. 564.

1907. **Overhead oscillating twin screw engines.**

1908. **Annular cylinder paddle engines.**

1909. **Overhead cylinder side-lever paddle engines.**
In addition to the above some special types are occasionally employed, as the Willan's three-cylinder plan for screw engines, See No. 592, also varieties of No. 593, high-speed types.

Section 104.—STRIKING AND HAMMERING: IMPACT.

The ordinary appliances for these purposes comprise hammers of all kinds and anvils or blocks of all shapes to suit the work, rammers and mallets of wood. The steam hammer being the machine almost invariably used, is too well known to need illustration. It is made with single or double standards, and though differing somewhat in details is practically the same machine wherever manufactured.

The following are apparatus employed for particular cases, and not so well known.

1910. **The drop hammer,** for power. The grip pulleys are put in gear by the hand lever to raise the hammer and shaft: it is sometimes worked by hand by a simple cord and pulley.

1911. **Dead blow power hammer.** The crescent-shaped crosshead bar has a positive motion from the crank pin, but the hammer head is attached to it by strong horizontal springs, and therefore has some little play above and below a horizontal line.

1912. **The pile engine and monkey.** The latter is generally raised by a hand or power winch, but a multiplying gear steam or hydraulic cylinder has been employed.

1913. **Another form of dead blow power hammer,** but with a straight laminated plate spring, to which the hammer head is fixed.

1914. **Another type of spring power hammer.**

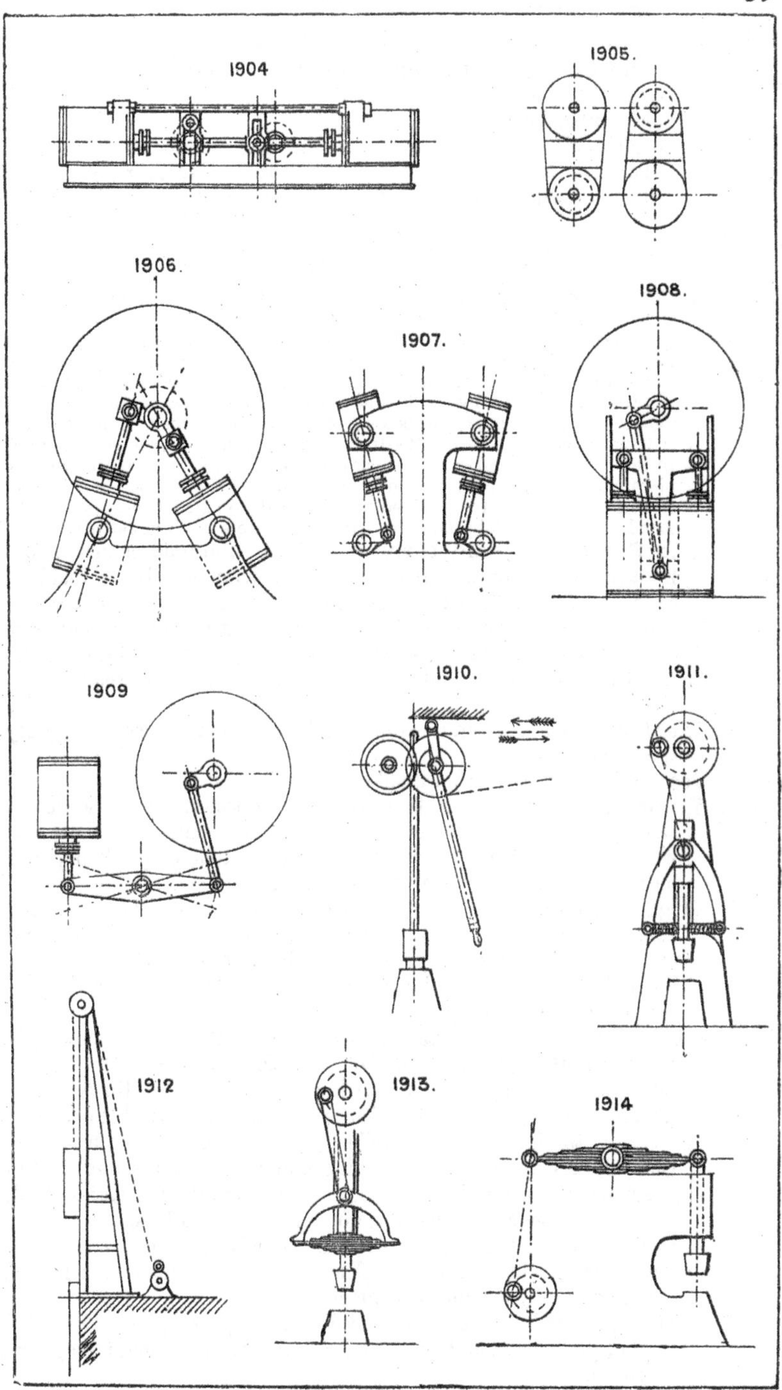
1904
1905.
1906.
1907.
1908.
1909
1910.
1911.
1912
1913.
1914

1915. **Revolving centrifugal rapid blow hammer.**

1916. **The old-fashioned tilt hammer**, still in use in many places, especially where water-power is employed.

For stamps, &c., see Nos. 250 & 271.

Besides the foregoing there is the gas hammer of Messrs. Tangye, the pneumatic hammer, and a variety of power hammer with variable stroke. See No. 1606.

Section 105.—SOUND.

Instruments for the production of sound are scarcely within the province of the mechanical engineer, but of late years several of them have been employed in connection with mechanical means for producing sound—for fog signals, whistling, and other forms of sound signalling. Musical sounds are produced by vibration of air from wind, string, or reed instruments. In wind instruments the vibration is produced by the lips and modified by the shape and length of the tube. Strings are either bowed, as in the fiddle, struck, as in the piano, or fingered, as in the harp. Reeds are springs vibrated by a current of air. In the harmonium and concertina class of instruments there are no tubes or pipes added to the reeds to modify the sounds produced; but in the organ pipe the reeds have pipes added which greatly augment and qualify the sounds. Other special sound-producing instruments are illustrated here.

1917. **The Siren, or steam turbine whistle**, the loudest instrument known, consists of a slotted cylindrical drum revolving inside a fixed drum; the slots are angular (see plan), so that the rush of steam revolves the inner loose drum rapidly and the sound is directed by the trumpet-shaped hood. A pair of slotted discs is also sometimes used for the same purpose instead of the slotted drums.

1918. **Mechanical fog-horn**; ordinary bellows are often used to supply the blast.

1919. **Iron gong**, struck with a muffled hammer.

1920. **Harmonium reeds, or free reeds**; the tongue covers a slot of same size and shape, and can vibrate into and out of it, but without touching its edges; the gravity of tone or pitch depends on the size and thickness of the tongue.

1921. **Organ reed pipe**; the tongue A in this case beats, with a rolling contact, upon the reed B, which is tubular, closed at bottom and opening at top into the pipe C, which extends upwards from the block D; E is the tuning wire which regulates the vibrating or free length of the tongue.

1922 & 1923. **Wood and metal organ pipes**, which are practically large whistles, the vibration of the column of air in the pipe being produced by the wind striking the edge of the lip A.

Steam whistles are bells with a ring-shaped slit below, from which the issuing steam strikes the lower edge of the bell.

Other forms are now made giving a more musical sound, and in some cases a double note, usually an interval of a major third, as C—E, by a modified form of pipe with two lips.

Striking bells of various shapes are extensively used.

Gongs are cheese-shaped metallic hollow suspended vessels, and are struck by a muffled hammer.

Musical sounds are also produced from slips of glass, or chilled iron, glass bells and tumblers, and also from resonant magnetic iron blocks.

1924, 1925, 1926. See Sec. 98.

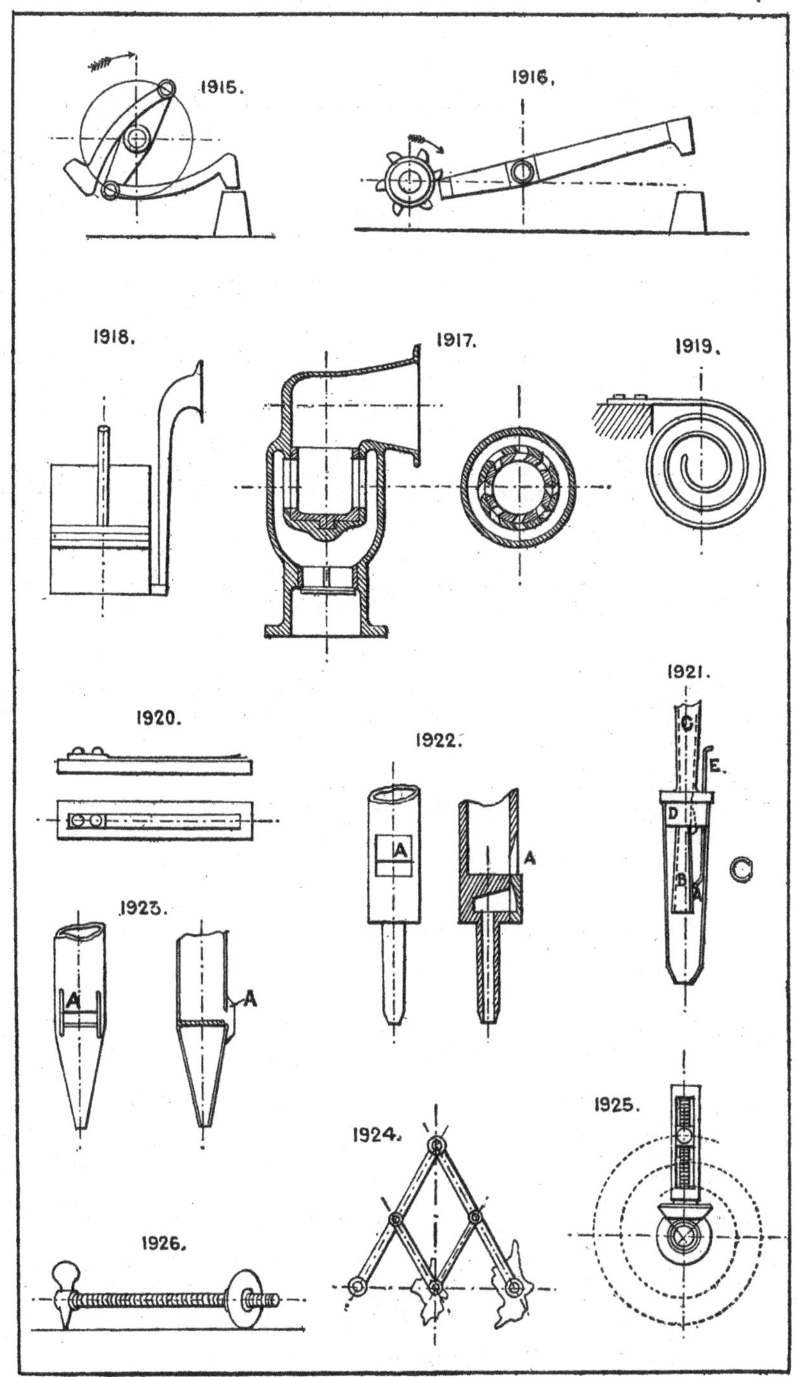
1915.
1916.
1918.
1917.
1919.
1921.
1920.
1922.
C
E.
D
A
A
B
A
1923.
A
A
1925.
1924.
1926.

Section 106.—DOORS, MANHOLES AND COVERS.

1927. **Application of a single crosshead and bolt** to close two covers, as in a pump clack-box.

1928. **Cone seated cover,** with hand-lifting crossbar and recess.

1929. **Crosshead and man- or mud-hole,** as commonly used for boilers, &c.

1930. **Cast-iron manhole and block;** T-head bolts are generally used, but also eye-bolts, as in No. 937.

1931. **Wrought-iron plate lid,** or cover for a tank.

1932. **Wrought-iron dished cover,** with hinged crossbar and T-screw used largely for gas retorts.

1933. **Furnace door;** hinged, with inside plate to protect the door from the heat.

1934. **Manhole door with water seal,** or packed recess to keep back gases, smell, &c.

1935. **Screwed plug handhole.**

1936. **Wrought-iron boiler manhole cover, and block,** a special manufacture.

1937. **Type of sliding door;** can be made airtight by planing the seatings.

Hinged doors are well-known. For hinging, see Section 50. For fastenings, see Locking Devices, Section 49.

See also Nos. 931, 937, 940, 962.

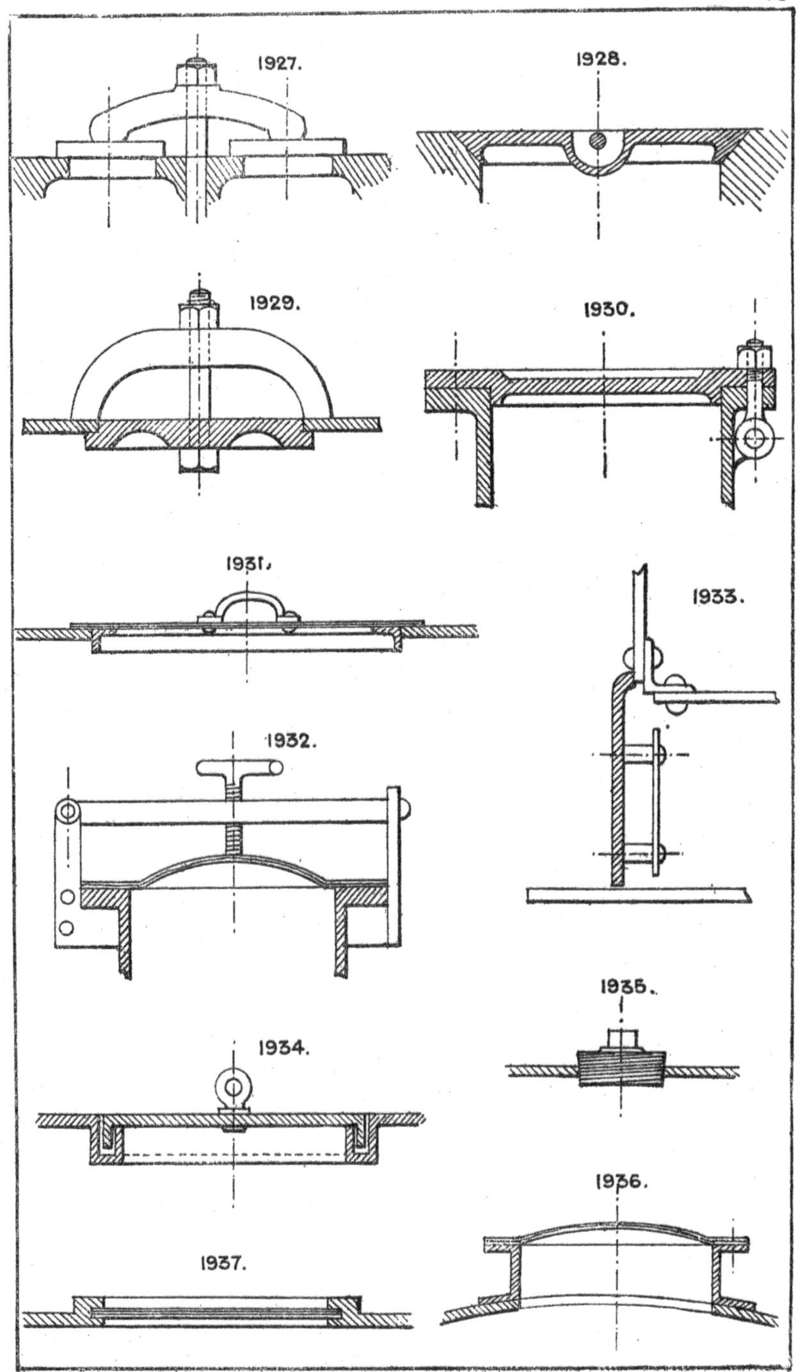
1927.
1928.
1929.
1930.
1931.
1933.
1932.
1935.
1934.
1936.
1937.

SAMPLE ADVERTISEMENTS FROM THE ORIGINAL EDITION

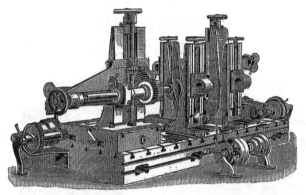

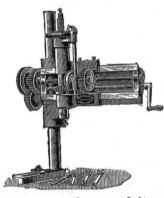

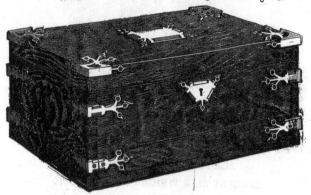

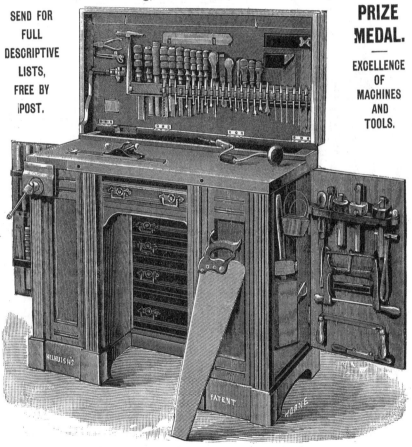

CHARLES CHU

Importers of AMERICAN

21, CROSS STREET,

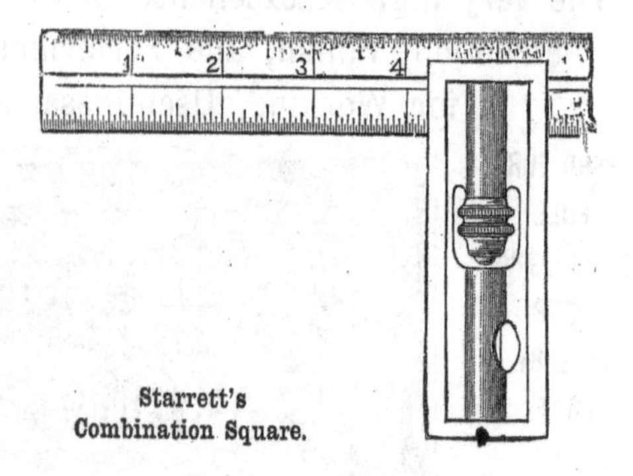

Starrett's
Combination Square.

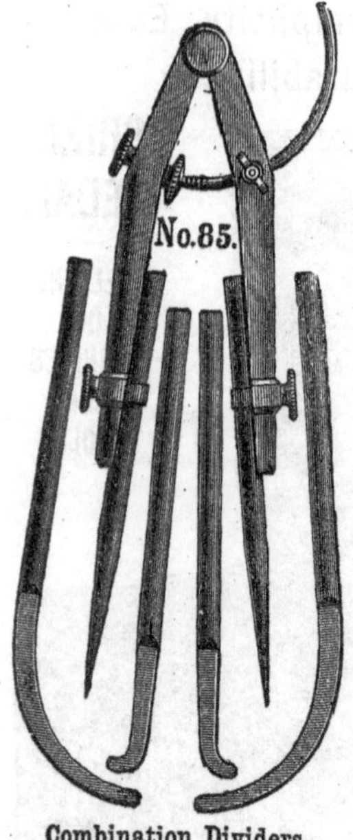

Combination Dividers.

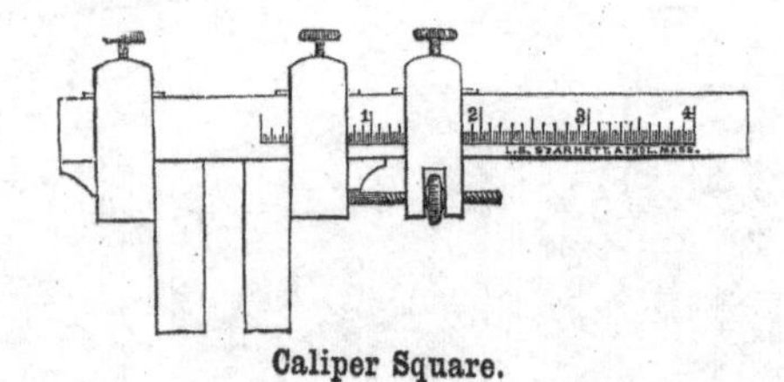

Caliper Square.

Stephens' Vices.

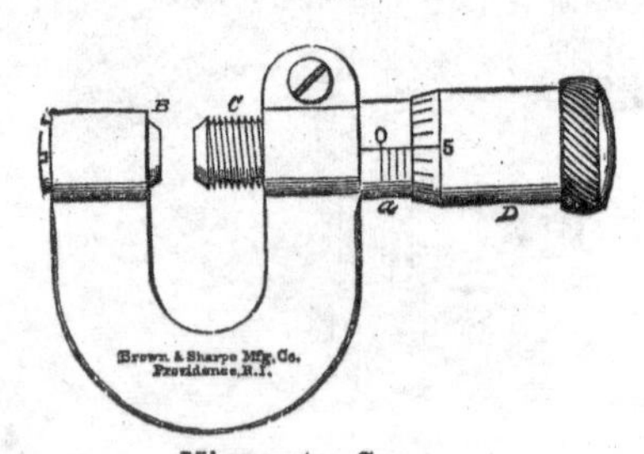

Micrometer Gauge.

We were the first to introduce AMERICAN Tools in Great Britain, and our twenty-five years' experience enables us to know the best and most suitable. We have the largest and most varied stock in the kingdom. We invite buyers to call and inspect the Tools in our Show Rooms.

We send our 1888–9 Catalogue, 160 pages quarto, over 600 Illustrations, for 1s.

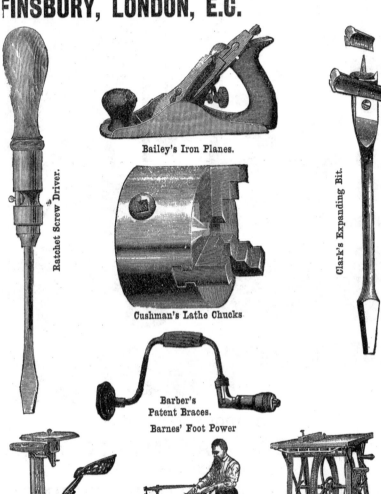

E. S. HINDLEY,

ENGINEER.

ECONOMICAL TO USE.

MODERATE IN PRICE.

LONDON SHOW ROOMS AND STORES—

11, QUEEN VICTORIA STREET, E.C.

WORKS—

BOURTON, DORSET.

STEAM ENGINES,

HORIZONTAL, VERTICAL, PORTABLE & FIXED,

IN ALL POWERS FROM ½ H.P. UPWARDS.

Suitable for Industrial or Electric Lighting Purposes.

These Engines are controlled by Mr. Hindley's new Governor, which maintains a perfectly uniform speed under sudden variations of work, and is adjustable when working, rendering them especially suitable for Electric Lighting Purposes.

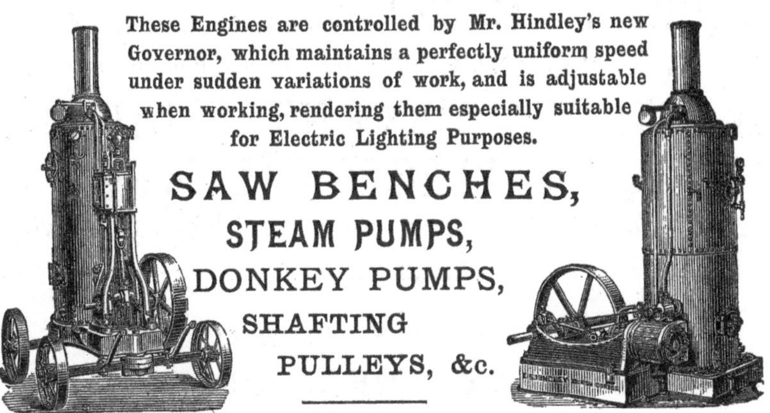

SAW BENCHES,

STEAM PUMPS,

DONKEY PUMPS,

SHAFTING

PULLEYS, &c.

Illustrated Catalogues, Photos, and Detailed Information Free on Application.

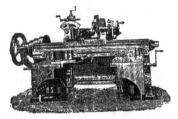

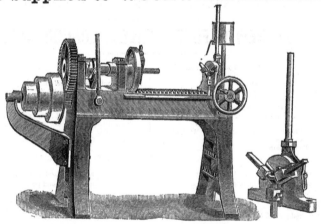